Studies in Economic Theory

*Editors*

Charalambos D. Aliprantis
Purdue University
Department of Economics
West Lafayette, IN 47907-2076
USA

Nicholas C. Yannelis
University of Illinois
Department of Economics
Champaign, IL 61820
USA

# Titles in the Series

M. A. Khan and N. C. Yannelis (Eds.)
Equilibrium Theory
in Infinite Dimensional Spaces

C. D. Aliprantis, K. C. Border
and W. A. J. Luxemburg (Eds.)
Positive Operators, Riesz Spaces,
and Economics

D. G. Saari
Geometry of Voting

C. D. Aliprantis and K. C. Border
Infinite Dimensional Analysis

J.-P. Aubin
Dynamic Economic Theory

M. Kurz (Ed.)
Endogenous Economic Fluctuations

J.-F. Laslier
Tournament Solutions and Majority Voting

A. Alkan, C. D. Aliprantis and N. C. Yannelis
(Eds.)
Theory and Applications

J. C. Moore
Mathematical Methods
for Economic Theory 1

J. C. Moore
Mathematical Methods
for Economic Theory 2

M. Majumdar, T. Mitra and K. Nishimura
Optimization and Chaos

K. K. Sieberg
Criminal Dilemmas

M. Florenzano and C. Le Van
Finite Dimensional Convexity
and Optimization

K. Vind
Independence, Additivity, Uncertainty

T. Cason and C. Noussair (Eds.)
Advances in Experimental Markets

F. Aleskerov and B. Monjardet
Utility Maximization. Choice and Preference

N. Schofield
Mathematical Methods in Economics
and Social Choice

C. D. Aliprantis, K. J. Arrow, P. Hammond,
F. Kubler, H.-M. Wu and N. C. Yannelis (Eds.)
Assets, Beliefs, and Equilibria
in Economic Dynamics

D. Glycopantis and N. C. Yannelis (Eds.)
Differential Information Economies

A. Citanna, J. Donaldson, H. M. Polemarchakis,
P. Siconolfi and S. E. Spear (Eds.)
Festschrift for David Cass (in preparation)

Mamoru Kaneko

# Game Theory and Mutual Misunderstanding

Scientific Dialogues in Five Acts

Translated by
Ruth Vanbaelen and Mamoru Kaneko

With 5 Figures, 11 Tables
and 12 Illustrations

Professor Mamoru Kaneko
University of Tsukuba
Institute of Policy and Planning Sciences
Ibrakai 305-8573
Japan
Email: kaneko@sk.tsukuba.ac.jp

This is the English translation of Kaneko M (2003) Game Theory
and Konnyaku Mondo (in Japanese). Nihon Hyoron-sha, Tokyo.

Cataloging-in-Publication Data applied for

A catalog record for this book is available from the Library of Congress.

Bibliographic information published by Die Deutsche Bibliothek
Die Deutsche Bibliothek lists this publication in the Deutsche Nationalbibliografie;
detailed bibliographic data available in the internet at *http://dnb.ddb.de*

ISBN 3-540-22295-2   Springer Berlin Heidelberg New York

This work is subject to copyright. All rights are reserved, whether the whole or part of the material is concerned, specifically the rights of translation, reprinting, reuse of illustrations, recitation, broadcasting, reproduction on microfilm or in any other way, and storage in data banks. Duplication of this publication or parts thereof is permitted only under the provisions of the German Copyright Law of September 9, 1965, in its current version, and permission for use must always be obtained from Springer-Verlag. Violations are liable for prosecution under the German Copyright Law.

Springer is a part of Springer Science+Business Media
springeronline.com

© Springer-Verlag Berlin Heidelberg 2005
Printed in Germany

The use of general descriptive names, registered names, trademarks, etc. in this publication does not imply, even in the absence of a specific statement, that such names are exempt from the relevant protective laws and regulations and therefore free for general use.

Cover design: Erich Kirchner, Heidelberg
Production: Helmut Petri
Printing: betz-druck

SPIN 11016953     Printed on acid-free paper – 42/3130 – 5 4 3 2 1 0

## Prologue

[The poet appears silently before the curtain]
A mere instance of thought leads to great achievements
Great plans and huge amounts of labor lead to scattered results

Only ten kilometers of opaque air make heat and water circulate
Another 200 kilometers higher the clear air drifts

Turbid seawater near the shore gives birth to fish
Transparent ocean water flows as a current

A mere moment in life has meaning
For the rest, people work consciously in a dream
[The poet leaves quietly]

[Morimori pokes his head through the curtain and speaks in an energetic voice]
I'm a graduate student at the Ph.D. program of economics at this university and I'm working with Professor Shinzuki on the field of game theory. Some people might think I'm only playing computer games or parlor games, because I said, "I'm working on game theory".

[Morimori looks at the audience]
Aha, some people think so. Actually, we play neither computer games nor parlor games. Game theory is the field of scientific study initiated in 1944 by von Neumann and Morgenstern, to analyze society and social phenomena mathematically[1]. Since society itself is very complicated, it is difficult to consider society directly in a mathematical way. So, these people started analyzing games, because games are regarded as prototypes of society and may be mathematically formulated. Therefore, Neumann and Morgenstern called their theory "the theory of games" rather than "game theory". Of course, their final goal is to explore social and economic prob-

---

[1] Von Neumann J, Morgenstern O (1944) Theory of games and economic behavior. Princeton University Press, Princeton.

lems, not to study computer games or parlor games.

[Morimori points out part of the audience]

Am I impressing you? You must be thinking I'm quite clever. To be honest, these aren't lines I made up; I'm only mimicking Professor Shinzuki when he starts his lectures on game theory. But he himself borrowed them from somewhere else, I think.

By the way, are you interested in the life of a graduate student? My job as a graduate student is to devote myself entirely to studying. For studying, you might think of taking classes, but for us, students after two years in the Ph.D. program, the course study is almost finished. We need to read research papers and do research. Our final goal is to write a Ph.D. dissertation. That's why we have to study hard and obtain good results. One of the requirements for a dissertation is to have a paper published in an academic journal. So, we are doing research day and night!

In Shinzuki's laboratory where I belong, Professor Shinzuki and Lecturer Hazamajime are always having discussions. Only recently I started to be able to participate in their discussions. Through discussions, I understand little by little how one does research. I hope soon I will be able to write a research paper.

[Looking behind the curtain, Morimori says in a small voice]

I have heard that some people had a lot of expectations for Professor Shinzuki when he was young. Now some people gossip, "Since the old days he was considered as a late blooming great talent but even the late blooming has passed without flowers and now he is totally forgotten." Just before, the poet said, "Great plans and huge amounts of labor lead to scattered results." He might be pointing at Professor Shinzuki.

Mr. Hazamajime is a very bright person, but we call him Mr. Majime, and I hear nowadays he has a good reputation in game theory and is very much in demand. Mr. Majime is always serious and doing research, but recently he seems to be worried about many things. Professor Shinzuki also seems to do research, but he has extremely regular hours. He comes to the university at the same time every morning and leaves at the same time every evening, while other professors often work until late at night. Sometimes I wonder if this person ever has any longings for success or worries about anything.

[Morimori, again in a loud voice]

This time Professor Shinzuki, Mr. Majime and I will have discussions on various subjects in economics and game theory. I'm very happy because at least I can participate in the discussions. Honestly speaking, I'm slightly anxious about whether I can follow their arguments. But, they seem happy to answer when I ask elementary questions. So I think somehow I will be able to take part.

Well, I hope you will stick around. It could be a lot of fun!

[From afar a loud voice is calling]

Is that Professor Shinzuki calling me? I think so. They might be about to start. Well everybody, please enjoy our discussions.

## The cast

Kurai Shinzuki: Professor in economics (all acts): Once promising but nowadays almost forgotten in his profession.

Toru Hazamajime: Lecturer in economics (Acts 1-5): Young economist who is successful and very much in demand nowadays. Everyone calls him "Majime", meaning "Serious".

Genki Morimori: Graduate student (Acts 1-5): Active and energetic Ph.D. student who has just started his research. Shinzuki and Hazamajime love him because of his childish character.

Show Hankawa: Lecturer at a university in Tokyo (Act 4).

Jan Hammer: Economist from Australia (Interlude 2)

Oliver Otsuki: Philosopher of science from Canada (Interlude 2)

K: The author

Poet: A masked prophet (Prologue, Acts 1, 3 and Epilogue)

Narrator: An anonymous economist (all acts)

## Setting

Shinzuki Laboratory: A laboratory in the social engineering institute at a university that was famous for its excellent campus and buildings two decades ago when the university was built (all acts except Interlude 2)

Mexican restaurant: Near the university (Interlude 2)

## Table of Contents

*Act 1  The reversal of particularity and generality in economics ................................................................. 1*
    Scene 1  The St Petersburg game ........................................ 1
    Scene 2  Paradox ................................................................. 10
    Scene 3  Generalities in economics and game theory ...... 20
    Scene 4  The reversal of particularity and generality ..... 27

*Act 2  Konnyaku Mondo and Game Theory ....................... 39*
    Scene 1  Konnyaku Mondo ................................................. 39
    Scene 2  The Prisoner's Dilemma and the Battle of the Sexes ..................................................................... 49
    Scene 3  Games with incomplete information .................. 58
    Scene 4  Rashomon ............................................................. 69

*Act 3  The market economy in a rage ................................ 79*
    Scene 1  Market equilibrium and social dilemma ............ 79
    Scene 2  Market failure and widespread externalities .... 88
    Scene 3  Epistemological consideration of perfect competition ........................................................... 99
    Scene 4  Institutional consideration of perfect competition ......................................................... 110

*Interlude 1  Clouds hanging over economics and game theory ............................................................ 121*

*Interlude 2  Game theory in a crisis ................................ 127*

*Act 4  Decision making and Nash equilibrium ................. 139*
    Scene 1  Recent topics ...................................................... 139
    Scene 2  Interpretations of the Nash equilibrium ........ 150

Scene 3 Infinite regress and common knowledge ......... 161
Scene 4 Mixed strategies ............................................... 174
Scene 5 Equilibrium as a stationary state ................... 189

*Act 5  The individual and society ........................................ 195*
Scene 1 Individualism .................................................. 195
Scene 2 Ism .................................................................. 207
Scene 3 Connections between the individual and society ............................................................ 217
Scene 4 Internal mental structure of the individual .... 226

*Epilogue ............................................................................. 241*

*Acknowledgements ............................................................. 245*

*The author ......................................................................... 247*

## Act 1  The reversal of particularity and generality in economics

Narrator: In this first act, Shinzuki, Majime and Morimori discuss about particularity and generality in economics and game theory. It is the distinctive characteristic of these fields that a player in a theory may think about that theory itself. If we take such a player's thinking into account, generality may become special, since the player's thought about a more general theory requires a stronger ability and thus becomes more special. Conversely, a more special theory allows a player to think less, and becomes more general than a general theory. This seems to be the meaning of "reversal" in the title of this act. The theme here is the consideration of a player's thought about the theory itself, which seems to be also the theme of the entire book.

### Scene 1  The St Petersburg game

[Shinzuki and Majime are talking in the laboratory. Morimori appears out of breath]

Morimori  Ah, Professor Shinzuki, I was looking for you. I have proven a terrific theorem!

Shinzuki  Mm, what kind of theorem is it?

Morimori  Do you remember the economic model I explained the other day? The basic set in the model was assumed to be finite but I have proven that the main theorem holds for an infinite set.

Shinzuki  It sounds good. What do you learn from your extension?

Morimori  I think it's evident. Since I eliminate the assumption that the basic set is finite, it is a true generalization.

Shinzuki  Mm, do you have any new concrete examples?

Morimori  Yes, of course, there must be an abundance of new concrete examples. Because I eliminated the assumption of finiteness, the range covered by the theorem has been broadened a lot. The contents of the theorem are much richer now. The elimination of the assumption makes the model more realistic because the assumption of finiteness is unrealistic.

Shinzuki   More realistic with an infinite extension?
Majime     Sir, I listened to Morimori's generalization. According to his explanation, it is indeed a generalization. If he writes a paper based on it, I think it could be published in *Journal of Theoretical Economics*. His generalization is more general than the generalization published in the latest issue. However, publication in the *Journal of Empirical Economics* could be difficult.
Shinzuki   It is publishable with reality as a selling point in *Journal of Theoretical Economics*. I wonder, though, where that reality is. Well, wherever it is, good for you, Morimori.
Morimori   No, no, the selling point is generalization, and reality is part of the generalization. Professor, you aren't pleased, though I have proven a new theorem. Mm ... why can't you be happy for me? You always tell us to try new things. I don't understand what you are thinking.
Shinzuki   Ah, one shall never know what another person is thinking.
Morimori   That is true but... Mr. Majime also says that if I write up a paper on my generalization, I can have it published in a first class

journal like *Journal of Theoretical Economics*. I agree with him. Anyway, it is more general than the generalization published in the latest issue of that journal. Professor, I want to ask about your criterion of what you call good research in economics or game theory, or what you call a novelty. Afterwards, could you please tell me if my generalization is interesting or not from your criterion?

Majime     I would like to know your criterion, too. So let's step away from Morimori's generalization for now. I would be very happy if you explain your general criterion on research.

Morimori     Mr. Majime agrees with me. The majority here asks you to explain your criterion. First, please explain it in detail, and then apply it to my theorem.

Shinzuki     Morimori, you are always tough and demanding! To argue a general criterion is a difficult task. It requires me to think about a general philosophy on economics and game theory. I have no confidence to build such a general philosophy. However, it may be challenging to think about it. Since a general philosophy is too abstract, it could be better to start with some particular examples. For now, let's take the St Petersburg paradox. You know this paradox, don't you?

Majime     Of course, I know it well. I start my lectures on expected utility theory with this paradox, and then I demonstrate the boundedness of derived von Neumann-Morgenstern utility functions using the argument of the St Petersburg paradox.

Morimori     I remember a little about what Mr. Majime said in his lectures, but not well. Let me recall, you toss a coin, and if it is heads, you would receive 2 cents. If it is tails...? It is something like this. It's boring, isn't it? Yes, yes, I recall then Mr. Majime asked whether some infinite series would converge or not. I recall this part well, because it was easy for me to calculate the sum of the series, hahaha. Overall, the discussions by Mr. Majime were too detailed, and the point was often unclear, so what he taught was quickly forgotten.

4  Scene 1  The St Petersburg game

Majime  My teaching is not the problem. Your understanding is the problem! You have a habit of blaming somebody else for your faults. You are an excellent student but you lack basic education.

Shinzuki  Stop it, stop it. It's better to think about the St Petersburg game now.

Majime  Okay, I will write the rules on the blackboard.

[Majime goes to the blackboard and draws Fig.1.1]

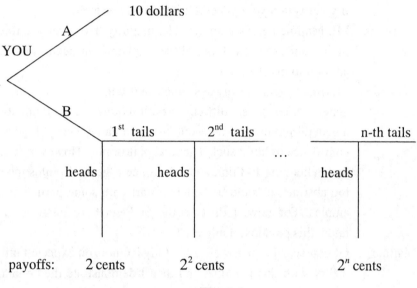

**Fig. 1.1.**

YOU are the decision maker, and YOU choose either $A$ or $B$. If YOU choose $A$, then YOU will receive a payoff of 10 dollars. If YOU choose $B$, then the payoff will be decided by coin tosses. First, YOU toss a coin. If it is heads, YOUR payoff will be 2 cents. If it is tails, YOU will toss it again. If it turns up heads on the second toss, YOUR payoff will be $2^2 = 4$ cents. If it is tails again, YOU toss it again. In general, suppose that heads come up on the n-th toss. Then, the payoff will be $2^n$ cents, and otherwise, YOU will go to the next toss.

## Act 1 The reversal of particularity and generality in economics

In sum, if YOU choose alternative $A$, the payoff will be 10 dollars, and if YOU choose alternative $B$, then the coin is tossed until heads come up, and if it happens on the n-th coin toss, the payoff would be $2^n$ cents. Morimori, do you remember the argument up to here?

Morimori    Yes, I'm starting to remember the St Petersburg game, and actually I understood your explanation better than during the lectures.

[Morimori comes towards the blackboard making Majime take a seat]

As I recall, the problem is whether YOU should choose $A$ or $B$. Since $B$ involves probabilities, it may be instructive to calculate the probabilities of some events. For example, the event of the payoff being 16 cents occurs if the coin tosses results in three successive tails followed by a head. The probability of this event is $1/16$. The probability of the payoff being higher than or equal to 16 cents is $1/8$, since the probability is equivalent to the probability of three successive tails.

What a boring game it is! If I were YOU, I would be sure to choose $A$. With that 10 dollars, I could go to Pizza Hut, have a medium pizza and a beer.

Shinzuki    Beer? I prefer to go to a bar.

Majime    Sir! Please don't interrupt us with a strange comment. Morimori, you should continue.

Morimori    Okay. The story continues as follows. I'll calculate the expected payoffs for alternative $A$ and $B$. We calculate them by multiplying the payoffs with their probabilities, and then taking the sum of those products. Since the payoff for $A$ is paid in dollars, while the payoff for $B$ is paid in cents, I'll convert dollars into cents. Is it okay, Mr. Majime?

Majime    Of course it is. Don't waste our time with such superficial questions.

Morimori    Sorry, Mr. Majime. Later, I will ask more profound questions. Now, the expected payoff for $A$ is $1000 \times 1 = 1000$ cents. On the other hand, the expected payoff for $B$ is calculated as follows:

$$2 \times \frac{1}{2} + 2^2 \times \frac{1}{2^2} + 2^3 \times \frac{1}{2^3} + \cdots = 1 + 1 + 1 + \cdots = +\infty \qquad (1.1)$$

Since infinity is bigger than $1000$, it is better to choose $B$. This is correct, isn't it, Mr. Majime?

Majime   Well done, Morimori. However, let me put it a little more academically.

When YOU assume the *expected payoff criterion* for YOUR decision making, YOU are supposed to choose alternative $B$. However, as you already said, if YOU look carefully at the probabilities of the resulting payoffs, $B$ is a cheap and poor alternative, and it is certain that you wouldn't choose $B$ if you are YOU. That's why this is a paradox.

Daniel Bernoulli insisted that the expected payoff criterion is inadequate, and instead, that we should substitute the *expected utility criterion* for it. He adopted $\log m$ as the utility function, and calculated the expected utility. This utility function is derived as a solution function of the differential equation expressing the Weber-Fechner law of physiology[1]. This law states, "When the stimulus is increased, the increment of sensation is proportional to the increment of stimulus but is reciprocally proportional to the absolute level of stimulus", and gives the utility function

$$u(m) = \log m.$$

---

[1] According to this law, the increment of utility $\Delta u$ is approximately proportional to $\Delta m / m$. When one takes the limit of this relation, one obtains the differential equation $du/dm = 1/m$, and $\log m$ is a solution function.

We adopt this utility function to calculate the expected utilities from alternatives $A$ and $B$, and then we obtain (1.2)[2].

$$A = \log 1000$$
$$B = \frac{1}{2}\log 2 + \frac{1}{2^2}\log 2^2 + \cdots + \frac{1}{2^n}\log 2^n + \cdots \quad (1.2)$$

Morimori, you were proud of your quick calculation of the series of $B$, right?

**Morimori**  Indeed, I was quick. Using a property of the logarithm for the series of $B$, we rewrite it as (1.3).

$$B = \frac{1}{2}\log 2 + \frac{2}{2^2}\log 2 + \cdots + \frac{n}{2^n}\log 2 + \cdots \quad (1.3)$$

Multiplying both sides by 1/2, we obtain formula (1.4), and then subtracting $B/2$ in formula (1.4) from B in (1.3), we reach formula (1.5). The right-hand side is a geometric series, and thus $B/2 = \log 2$. We have $B = 2\log 2$ or $B = \log 4$.

$$\frac{1}{2}B = \frac{1}{2^2}\log 2 + \frac{2}{2^3}\log 2 + \cdots + \frac{n}{2^{n+1}}\log 2 + \cdots \quad (1.4)$$

$$\frac{1}{2}B = \frac{1}{2^1}\log 2 + \frac{1}{2^2}\log 2 + \cdots + \frac{1}{2^n}\log 2 + \cdots \quad (1.5)$$

**Majime**  Thank you very much. Thus, according to the expected utility criterion, YOU compare $\log 1000$ with $\log 4$, and should choose alternative $A$. Here, we have reached one solution of

---

[2] Bernoulli D (1954, original 1738) Exposition of a new theory on the risk. Translated by Sommer L, Econometrica 22: 23-26.

the paradox by changing the expected payoff criterion to the expected utility criterion.

Nonetheless, the above argument doesn't completely extinguish the paradox. With the right choice of the payoffs in alternative $B$, we can make $B$'s expected utility infinite once more. Thus, as a conclusion we need to develop the general theory of expected utility.

Morimori   Now I remember the entire story well. To make $B$'s expected utility once again infinite, we have to change the payoff from $2^n$ to $2^{2^n} = 2^{(2^n)}$ for heads on the $n$-th coin toss. I understand the explanation up to here. However, I didn't understand your last conclusion, that is, "Thus, we need to develop the general theory of expected utility"[3].

Although you started your lectures with the expected payoff criterion and discussed the expected utility criterion, the rest of the lectures became about mathematical conditions such as transitivity, continuity, independence etc. Then you went to the mathematical derivation of a numerical utility function from a preference relation. You never returned to either the expected payoff criterion or the expected utility criterion. I wondered what happened with those criteria in the theory of expected utility.

Majime   Yes, you are right. The main topic of expected utility theory is to analyze mathematical conditions for a preference relation to have a numerical utility function. A decision criterion exists in the mind of the decision maker, and it is expressed as a numerical utility function when certain conditions are fulfilled. Transitivity, as well as other conditions for a preference relation, is natural and plausible when we consider a natural decision criterion, and this is why expected utility theory is useful.

---

[3] For expected utility theory, see Hammond P (1998) Objective expected utility: a consequential perspective. In: Barbera S et al. (eds) Handbook of utility theory Vol.1. Chap.5 pp 143-211. Kluwer Academic Press, Amsterdam.

Morimori   Mr. Majime, what you said after "this is why" doesn't follow your previous statement. One possible way to continue your sentence is: "this is why expected utility theory didn't discuss which decision criterion should be chosen".

Majime   Mm... you're right.

[Morimori is quiet for a little while]

Morimori   Hey, I think expected utility theory doesn't answer the St Petersburg paradox at all. The answer to the paradox changes according to the criterion. But the original question of which criterion we should adopt remains.

Majime   Indeed, expected utility theory doesn't talk about which criterion should be used. However, at least, the theory does claim to take the expected value of utilities.

[Shinzuki is looking at his watch several times]

Shinzuki   Ah... it is about time for me to go home. Sorry. Today's dinner is, unless I'm mistaken, tofu and something else. My wife told me "buy some tofu on your way back".

When you watch a police series on TV, investigators always say, "when you get stuck in an investigation, you should go back to the crime scene". In our case, the crime scene is why the St Petersburg paradox is a paradox. I shall listen to the rest tomorrow, so please think it over a bit more.

[Shinzuki hurries and leaves]

Morimori   If the St Petersburg paradox is a crime, then Daniel Bernoulli was a criminal, wasn't he?

Majime   Professor Shinzuki only used that phrase as an analogy. You shouldn't take it so seriously.

Morimori   Yes, I know. But how is this related to my generalization theorem?

Majime   There'll be plenty of time tomorrow. By the way, Professor Shinzuki always goes home at this time. After dinner, he's in bed by ten o'clock. He's such a child.

## Scene 2  Paradox

[The next day, the three of them are having coffee in the laboratory]

Shinzuki   Yesterday, where did we leave our discussion?

Morimori   Mr. Majime explained the St Petersburg game. Then, Professor, you ran out of time and left us with the phrase "go back to the crime scene when you get stuck".

Majime   I followed your advice, and I thought about why the St Petersburg paradox is a paradox.

According to some book, paradoxes are broadly categorized into two kinds. The first one is called genuine paradoxes. These occur when contradictory propositions can be derived from one claim or one theory. Nevertheless, the claim or the theory should be seen as coherent from a normal point of view, and no contradiction is expected. A genuine paradox is also called a logical paradox. The Cretan Liar[4] is one example and the paradoxes in the naive set theory are others.

Morimori   What is the other kind?

Majime   The second kind is called a pseudo paradox, in which case two or more criteria or axioms contradict one another, though each is normally accepted. In the case of the St Petersburg paradox, no contradictions are derived from the St Petersburg game itself. Instead, a contradiction lies between the decision maker's own decision and the recommendation given by the expected payoff criterion. It is a paradox only in this sense. Therefore, this is a pseudo paradox but not a genuine paradox.

Additionally in both kinds, to be called a paradox requires not only a contradiction, but also an indication of some problem in our normal thought process that is regarded as correct.

[Shinzuki, having listened in admiration, suddenly becomes authoritative and sarcastic]

Shinzuki   According to your classification, we can call Arrow's general impossibility theorem a pseudo paradox, can't we? Arrow as-

---

[4] It is often called the "liar paradox". See Oxford companion to philosophy, p.483. (1995), Oxford University Press, Oxford.

Act 1  The reversal of particularity and generality in economics  11

sumed five axioms on a social welfare function, each of which seems plausible, and then proved that these five axioms are inconsistent. It was paradoxical when Arrow proved the theorem, and it remains a paradox for some people.

After Arrow, many other impossibility theorems were proved within the axiomatic systems that were constructed inconsistently in the first place. Even then, some people have continued such research while learning nothing from Arrow's pseudo paradox, and blindly repeating their efforts of failure.

[Majime, slightly upset]

Majime   Sir, you always become so negative the moment you start talking about something. If I'm not mistaken, Arrow's general impossibility theorem is one of the reasons for his Nobel Prize[5].

Besides, being so negative makes you unable to write research papers. That is why people in our profession look at you skeptically.

---

[5] For Arrow's impossibility theorem, see Arrow KJ (1951) Social choice and individual value. Yale University Press, New Haven. More detailed explanation and evaluation are given in Luce RD, Raiffa H (1957) Games and decisions. John Wiley and Sons, New York.

Morimori   I ... I have something positive to ask. Is it the difference between a genuine and a pseudo paradox that one is genuine and the other is pseudo?

Majime   Hahaha, it is a tautology and doesn't explain the difference. Mm, to put the difference clearly, wait a moment, what really is the difference?!

Shinzuki   A profound question! It must be impossible to genuinely distinguish between genuine and pseudo. Such a distinction would be no more than a pseudo distinction and the distinguisher would be no more genuine than a pseudo scholar. Whether genuine or pseudo, if you understand the reason for the paradox, then it stops being paradoxical. In this sense the distinction between genuine and pseudo is just a matter of degree.

Majime   Is that all?

Shinzuki   I think that even for the famous Cretan Liar, the paradox disappears if one defines rigorously the language to be used. But since that is not a simple task, let's take a simpler example, say, Arrow's theorem.

The essence of Arrow's theorem lies in the paradox of majority decision making. Suppose that there are three people, 1, 2 and 3, with three social alternatives $x, y$ and $z$. The players decide, by majority voting, to choose one of these three social alternatives. Their preference relations are given as follows:

$$\text{Player 1: } xP_1yP_1z,$$
$$\text{player 2: } yP_2zP_2x,$$
$$\text{player 3: } zP_3xP_3y.$$

In other words, player 1 prefers $x$ to $y$ and $y$ to $z$, player 2 has his preferences in the order of $y, z, x$, and similarly, player 3 $z, x, y$.

First, consider the social choice from $x$ and $y$. Since players 1 and 3 prefer $x$ to $y$, $x$ is chosen over $y$ by the majority. In the comparison $y$ with $z$, since 1 and 2 prefer $y$ to $z$, $y$ is chosen over $z$ by the majority. Finally, when $z$ is compared with $x$, the majority chooses $z$. Thus we have a cycle, which is depicted in Fig.1.2 and is called the Condorcet cycle.

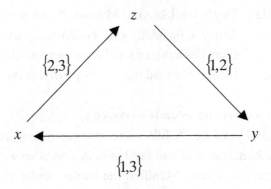

**Fig. 1.2.**

Morimori  So, Professor, where is the paradox in this voting situation?

Shinzuki  That is the problem. Mm, you have to ask the question, "Why is it a paradox when a cycle of social decisions appears?" This cycle claims that the majority voting may not define the best choice. However, a lot of people have the belief that majority voting, i.e., democracy, leads always to the best social alternative that satisfies everybody. The above cycle contradicts this belief. Therefore, the above cycle is a paradox for people who are obsessed with that belief.

It is not a paradox at all for people who accept the fact that democracy doesn't necessarily define the best social alternative or that it may often lead to chaos. Also, for people facing the above example and finding their way of thinking too naive and revising their beliefs, the paradox is no longer a paradox.

Morimori, logically speaking, what kind of people remain?

Morimori  Mm ... remaining are people who saw the Condorcet cycle but continued to believe that democracy always leads to the best social alternative.

Majime  Sir, you led Morimori to a bad conclusion. You should behave like a university professor.

## Scene 2 Paradox

Shinzuki    I'm sorry. I will make sure to state the bad conclusions myself from now on.

Majime    I understand the voting paradox. But, what do you think about Arrow's theorem?

Shinzuki    That's not like you, Majime. If you read carefully the proof of Arrow's theorem, you should understand that the Condorcet cycle has more or less the essence of Arrow's theorem.

Majime    I see. I will read the proof of Arrow's theorem more carefully.

[The poet appears suddenly on the stage]
    2,500 years ago Achilles started to compete with the tortoise
    Achilles ran hard and fast but didn't catch up with the tortoise
    Everybody sang "Achilles, run harder, harder and faster"
    Achilles ran harder and faster, and passed the tortoise taking a siesta
[The poet leaves quietly]

Morimori    What is that all about?

Shinzuki    I think the poet also wants to solve a paradox. But he mixed up Zeno's paradox with Aesop's story.
    Well, let's return to the St Petersburg paradox.

Majime    All right, following the discussions today, we should consider in what sense the St Petersburg paradox is really a paradox.
    The expected payoff criterion is quite often used when probabilistic elements are involved. However, in the St Petersburg game, a person with sufficient imagination would clearly choose the alternative rejected by the expected payoff criterion. This paradox tells us that the expected payoff criterion may not express our own decision making. Therefore, we need expected utility theory.

Morimori    Wait a moment. Isn't that the same as yesterday's discussion? Expected utility theory doesn't discuss decision criteria but still you want to go to expected utility theory.

Shinzuki    You are right. We shouldn't jump from the St Petersburg paradox to expected utility theory. Rather, let's consider how we should look at this paradox.

After Daniel Bernoulli, many people tried to find a resolution to the paradox while staying within the scope of the expected payoff criterion. For example, people ignore, react slightly or too much to very small probabilities. Or they cannot imagine very big payoffs. When one incorporates these factors, the result differs significantly.

Lloyd Shapley gave a clear-cut solution, claiming that we should take into account the budget constraint of the bookmaker of the St Petersburg game[6]. In fact, this is related to our original problem of how Morimori's theorem is evaluated.

Morimori  I'm happy to hear that. But why do we need the bookmaker in the St Petersburg game?

Majime  Actually, I modified the original St Petersburg game slightly in order to consider it directly in expected utility theory. The original form consists only of alternative $B$, and YOU are asked how much YOU may pay for the participation in $B$. Since the expected payoff from $B$ is infinite, YOU participate in the game whatever the price for $B$ is. So, this is why the bookmaker is involved in the game.

Morimori  I see. Please go on.

Shinzuki  Let's return to the St Petersburg game described on the blackboard, and consider the budget constraint of the bookmaker.

If YOU were very lucky in the coin toss, and the first head came up on the $100^{th}$ toss, then YOUR payoff would be $2^{100}$ cents. This number is perhaps bigger than the number of molecules in one square centimeter of an ideal gas. Morimori, do you know this number?

Morimori  Are you asking about $2^{100}$, or about the number of molecules in gas?

Shinzuki  Of course, the latter.

Morimori  Perhaps, I learned it in my chemistry classes at senior high.

---

[6] Shapley L (1977) The St. Petersburg paradox: a con game. Journal of Economic Theory 14: 439-442.

Majime  The number of molecules in one mole of an ideal gas under atmospheric pressure 1 and $0°C$ is called Avocado's number, right? That was about $6 \times 10^{23}$. The volume of 1 mole of an ideal gas is about 22 liters, so the number of molecules in 1 square centimeter of gas is, mm, about $10^{20}$.

Shinzuki  It's not Avocado's number but Avogadro's number, Majime. It is quite rare for you to make such a mistake. According to the picture in the textbook of chemistry, Dr. Avogadro looks like a mad scientist with his head of the shape of an upside down triangle. Some famous economist has a head similar to Dr. Avogadro.

Majime  Who is that? Well, we shouldn't talk about mad scientists. We are economists, so let's consider economic figures. Since we are talking about budget constraints, we can use the Japanese national budget. Morimori, do you know what the annual budget of the Japanese government is this year?

Morimori  Mm, the Japanese population is about 125 million, and with 1,000 dollars/person the total amount becomes about 125 billion dollars[7]. But will everyone pay 1,000 dollars, even children, elderly, homemakers and homeless?

[Majime is disappointed]

Majime  Are you really studying economics? The Japanese national budget of this year is about 800 billion dollars. Since your detour will take too much time, I should continue by myself.

First, we need to use cents rather than dollars in the St Petersburg game, 800 billion dollars is 80 trillion cents, i.e., $80 \times 10^{12}$ cents. Now, using a calculator, $80 \times 10^{12}$ is, uhm..., somewhere between $2^{46}$ and $2^{47}$. Suppose that the bookmaker of the St Petersburg game has a budget constraint with this amount. As in the previous case where we go to the 100th coin toss and it is

---

[7] In June 2004, the exchange rate between US dollar and Japanese yen is about 110 yen to the dollar. Hence, 1 yen is about 1 cent, and 1000 yen is about 10 dollars. To simplify the calculation, when discussing the Japanese national budget, it is assumed that 1 dollar = 100 yen.

heads, the payoff is also $80 \times 10^{12}$ cents. If the coin toss goes to more than $46^{th}$ toss, the payoff would be always $80 \times 10^{12}$ cents. This is illustrated in Fig.1.3.

Shinzuki   Hm, you mean that the entire Japanese population supports that bookmaker. In the worst case, each citizen will pay, mm ... about 7,000 dollars to him. The story is getting realistic now.

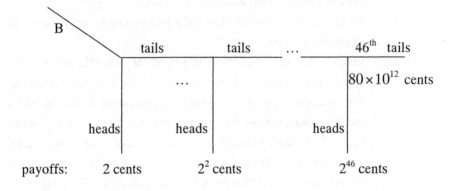

Fig. 1.3.

Majime   Let me calculate the expected payoff from this modified alternative $B$. Since 1 is added 46 times and plus at most 1, it becomes $1+1+...+1 = 46+1 = 47$. This means that the expected payoff for $B$ is about 47 cents and compared to the 10 dollars expected payoff for $A$, you should choose alternative $A$. Aha, with this the St Petersburg paradox is happily solved. Good.

[Shinzuki gets upset with the unexpected conclusion]

Shinzuki   You did the correct calculation, but is your final conclusion satisfactory for you? For example, it is an implication that your favorite theory, expected utility theory, becomes unnecessary. It is not a good idea to jump to a conclusion so quickly. Let's discuss the problem more carefully.

Majime   Okay, it is natural to introduce a budget constraint, forgetting the assumption that it is the Japanese annual budget. Without this introduction, the original alternative $B$ had an infinite

|  |  |
|---|---|
|  | number of possible payoffs. On the other hand, with the introduction of a budget, the set of payoffs is restricted to a finite set. Is the St Petersburg paradox caused by the fact that the original alternative $B$ involves an infinite number of payoffs? |
| Shinzuki | Well, it would be difficult to think about a general theory of infinities, so let's limit ourselves to only the facts we have seen. Here, I emphasize the fact that we extended the St Petersburg game by introducing a budget constraint. |
| Morimori | Mm... does the introduction of a budget constraint become an extension of the problem? |
| Shinzuki | Indeed, this is the point. The set of the possible payoffs becomes more restrictive but the theory itself is now extended. For example, unless the theory of consumer behavior has a budget constraint as its component, the theory won't make sense. Only with a budget constraint we can treat it as a meaningful problem. Thus, the introduction of a budget constraint is an extension of the theory in the sense that an additional structure becomes newly available. |
| Majime | Sir, now I'm starting to understand your plan of the present discourse. You seem to be saying that there is no clear-cut relation between "finite" and "infinite" with respect to generality, right? From there you want to bring us to the conclusion that Morimori's generalization theorem of extending the finite basic set to an infinite one is not that interesting. |
| Shinzuki | No, no, I had no intention at all to lead to such a conclusion and even now I don't think in that way. We have just followed a logical discourse. In addition, the logic was driven by the two of you, each of whom is regarded as superb by other people and more importantly by yourselves. |
| [Majime is unwilling to accept but...] | |
| Majime | All right, I'll try to summarize our discussion. When the set of payoffs is infinite in the St Petersburg game, some paradox might appear. However, when the problem is extended with the introduction of a budget constraint, the paradox could be solved. |

Morimori  Mathematically speaking, its introduction makes the set of payoffs finite. I think, Mr. Majime, that your lectures on expected utility theory start from the case where the set of alternatives is finite. Then, in order to handle the St Petersburg game you extend it to the case where the set of alternatives is infinite[8]. This infinite extension becomes unnecessary, according to the previous conclusion.

Mm... I still feel that the finite restriction of the set of payoffs sounds more like particularization than generalization. But I'm also starting to understand your plan, Professor.

[Shinzuki looks happy and says slowly]

Shinzuki  Both of you think I planned from the beginning to lead you to that conclusion, but my actual intent differs slightly. When I heard Morimori's generalization, a little voice in my head said, "it sounds a little strange and you, Kurai, should pursue this". Thus I chose the St Petersburg game for this pursuit, and up to now my choice seems to have been appropriate.

Majime  Okay, okay, Sir, your intention is now clear to me. Morimori, shall I tell you how he wants to carry out his pursuit?

[Majime speaks with his head up and his voice in a tone of an actor]

> *"The ultimate goal for a scholar is to approach the truth. However, approaching the light of the truth is a blinding experience and often not beneficial for oneself. A lot of scholars stop their quests for truth in such a case, but the most praiseworthy behavior for a scholar is not to stop, no matter what one's quest might cause to oneself.*
>
> *That is just like Oedipus, following the oracle and foreboding that he would smash his own eyes, he couldn't stop exposing*

---

[8] It is necessary not only to derive a utility function representing a preference relation, but also to show the fact that the form of the derived utility function takes expectation of utility values. Several books discuss the case when the set of nonprobabilistic (pure) alternatives is finite. Some axioms need to be added to obtain the expected utility theorem when the set becomes infinite. In that sense it is necessary to extend the axiomatic system.

*the fact that he killed his own father and that his wife is his own mother."*

Sir, you felt yourself to be Oedipus[9], and following your premonition, you pursued the truth.

[Morimori is impressed]

Morimori  Mr. Majime, that is how a scholar should be. I admire it.

Majime  I was also impressed when I heard these words once from Professor Shinzuki. However, those fishy lines reek of someone else's breath.

Shinzuki  Yes, they are, but no, I should be proud that some combinations are my own. Well, Mr. Majime, you saw right through me.

It is almost noon. Shall we go for lunch? We will continue our discussion in the afternoon.

[The three leave, being quite satisfied]

## Scene 3  Generalities in economics and game theory

[The three return from lunch, appearing sleepy on the stage]

Morimori  From the discussions of this morning I came to understand there is a complicated relation between particularity and generality. But our emphasis shouldn't be on the particular example of the St Petersburg game. Instead, I want to know about the general criterion of what generality is and what criterion is used to distinguish between particularity and generality. Without knowing these, I can't evaluate my generalization theorem. Professor, could you please explain your general theory of generalities?

[Majime says challenging]

Majime  Imitating your way, Sir, I should put a constraint on the way you will continue your discussion.

---

[9] Sophocles: Oedipus the king. Translated by Storr F (1912). Harvard University Press, Cambridge.

|          | First, you should talk about a general theory rather than examples and analogies. Next, you shouldn't use us as an engine to develop your theory. In other words, you should talk about your general theory of generalities in your own language. |
|---|---|
| Shinzuki | Th... that is harsh! Mm, but since I'm in the minority, I should follow the decision of the majority. Bu... but when I derail or when I get stuck, please help me, won't you? And, please forgive me if I, once in a while, use an example or an analogy. |
| Morimori | It has always been like this! But it's fine, isn't it? Mr. Majime. |

| Majime | Okay. Fine with me. |
|---|---|
| Shinzuki | Well, I'll try my best. |
| | To explain generally what generality is, I will take the position of the present-day mathematical logic. It is based on the comparison of two axiomatic systems and of theorems in these systems. To talk about axiomatic systems rigorously from this viewpoint, we have to formulate language, inference rules and logical axioms. We would need one semester to discuss them properly, but here I will cheat a little. |

|          | |
|----------|---|
|          | First, suppose that an axiomatic system $A = (A_1,...,A_n)$ is an extension of another axiomatic system $B = (B_1,...,B_m)$. This means that any statement in $B$'s language can be expressed in $A$'s language and furthermore, that any statement provable in $B$ is also provable in $A$. This notion of extension allows for a formal comparison between axiomatic systems[10]. |
| Morimori | Is this explanation applicable to my generalization theorem? |
| Shinzuki | No, I talked about the comparison of two axiomatic systems, in other words, the generality of two theories rather than two theorems. |
| Morimori | Does generality differ between theories and theorems? |
| Shinzuki | Indeed. A "theory" is synonymous with an "axiomatic system". My contention is that generality for a theory is entirely the reversal of that for a theorem. |
| Morimori | Is that true? |
| Shinzuki | Yes, it is. I myself was very surprised when I noticed it. Let me see in retrospect when I started to think about this reversal. It was more or less two decades ago when I started working at this university. At that time, the large campus and new buildings of this university were quite famous in Japan. However, buildings were constructed everywhere, and it was so dusty. Hm ..., where am I heading? |
| Majime   | Sir! I would be most happy to listen to the content of your intention of reversal if you could only keep to the point. |
| Shinzuki | That is right! I would also feel very content if I succeed in explaining the content of my contention to you. |
|          | In your case, Morimori, it is about generality for a theorem and you should think about it in the following way. One statement $T$ is proved from certain axioms. For example, let us suppose that $T$ is proved from axiomatic system $A = (A_1,...,A_n)$, as well as that $T$ is provable from the weaker system $B =$ |

---

[10] For explanations of inference rules, logical axioms, mathematical axioms, see Mendelson E (1987, 3$^{rd}$ ed) Introduction to mathematical logic. Wadsworth & Brooks, Belmont.

|          | $(B_1,...,B_m)$. That is, $T$ is proved under a weaker assumption in a theory. In this case, however, the applicability of $T$ is broadened.<br>Do you understand? |
|----------|---|
| Morimori | Yes, more or less. |
| Shinzuki | In fact, we can compare $A = (A_1,...,A_n)$ with $B = (B_1,...,B_m)$ in another manner. A more detailed discussion is possible in the extended axiomatic system $A = (A_1,...,A_n)$. In $A$, we can derive more conclusions than in $B$ and in this sense $A$ is richer than $B$. I would like to use the word "generality" for the case of this comparison. That is, $A$ is a generalization of $B$. For example, Albert Einstein's general theory of relativity is a generalization of his special theory of relativity. |
| Majime   | Wait a minute. It sounds different from standard practices in mathematics. We normally say that the theory of topological spaces is a generalization of the theory of metric spaces. According to your claim, the theory of metric spaces is a generalization of that of topological spaces. However, this isn't true in mathematics, is it? |
| Shinzuki | That is correct. Following conventions in mathematics, the word "generalization" doesn't mean an extension of a theory but often it means a generalization of a theorem. The reason why the theory of topological spaces is a generalization of that of metric spaces is that the theorems shared are proved from weaker axioms in the theory of topological spaces. Less can be stated in a weaker theory but the domain of applications of each shared theorem is broader[11]. |
| Majime   | Sir, which do you call more general, the more detailed axiomatic system $A = (A_1,...,A_n)$ or the weaker axiomatic system $B = (B_1,...,B_m)$? |
| Shinzuki | I'm interested neither in making an axiom system weaker nor in proving a theorem already proved by somebody else. It is suffi- |

---

[11] Cf. Royden HL (1963) Real analysis. MacMillan Publishing Co, London.

cient to make such a generalization when necessary. Personally I'm interested in a more detailed theory, including an extension of language. That is what I would like to call a generalization. For now, however, let's follow the standard convention in mathematics.

Morimori  Thank you very much for your explanations, but it sounds as if my generalization is without much value.

Shinzuki  You don't have to be so pessimistic. It is a generalization of the theorem, indeed.

Until now I talked purely about the generality of an axiomatic system as well as that of a theorem. However, when we start talking about economics or game theory, the problem becomes more delicate. In fact, two new problems arise. First, mathematical economics and game theory are mathematical theories dealing with social phenomena.

In such a case, the choice of an axiomatic system becomes a problem of social science and not just mathematics. A mathematical theory in social science is not merely about extant mathematical objects, but includes choices of axioms as important components of a theory. Accordingly, it is a mistake for a mathematical social scientist to specialize in a given mathematical theory. He should consider also the background of why a certain axiomatic system was chosen[12].

[Shinzuki is elevating his voice, with his spit flying]

Morimori  Professor, please don't get too excited.

Shinzuki  Mm, I shall be careful. The other problem is to include a thinking and calculating agent in the system, namely a player in the terminology of game theory. We, game theorists or economists, often assume that a player has a strong reasoning and inference

---

[12] The first chapter of von Neumann J, Morgenstern O (1944) The Theory of game and economic behavior. Princeton University Press, Princeton, discussed in detail what one has to take into consideration when one wants to develop a mathematical science (especially theoretical economics), comparing the historical development of physics. The first chapter provides a philosophy for development of mathematical theories of social and economic sciences.

## Act 1 The reversal of particularity and generality in economics

ability. However, this reasoning ability of a player is not explicitly formulated.

Indeed, a skilled game theorist almost always starts with some discussion of this part. But often this is discussed only among game theorists. In recent textbooks, only the mathematically formulated part is clear but the informal part remains obscure. Consequently, beginners in game theory, or people with no contacts with game theory circles can hardly become familiar with this part.

Morimori  Do you mean game theorists use a lingo understandable only in their own closed circle and show only mathematically formulated results to outsiders?

Shinzuki  Yes, that is exactly what I mean.

Morimori  Isn't that the same as a new religious sect? That's terrible. But Professor, aren't you a game theorist?

Shinzuki  Well, something like that. Let me continue. Intrinsic difficulties are often hidden in economics and game theory. This is inevitable since there are too many things that remain implicit and are not even consciously thought about.

It is rare for a circle to try consciously to clarify their jargon. Consequently, people coming from different circles are likely to misunderstand the terminology of other circles.

Morimori  It sounds awful. I don't want to become such a game theorist. Can you please return to the problem of generality?

Shinzuki  Sure. The problem is that many things are left behind in the background of a mathematical formulation and they are not discussed consciously. Especially the player's knowledge or rationality is left untouched, but this is an important part in the theory.

In game theory papers you often see, "We assume here that the rules of the game are common knowledge". This is a good, no, a bad example of how mathematically formulated components and unformulated components are mixed.

Finally I have reached what I wanted to talk about.

Morimori  Finally we can hear your true intention!

Shinzuki  Mm ... I'm not sure about my own true intention, but what I want to say at least is the following. Suppose that a certain part of the mathematically formulated theory is generalized. As I said before, economics and game theory include a player and usually his knowledge or reason is not included in the formalized part[13].

Now, assume that the generalized part is directly related to the player's thoughts. In this case, by this generalization, a player needs to think about more possibilities or needs to calculate more. Perhaps, on the contrary, he could decide on his own behavior through a simple calculation before generalization.

If we assume that the player can act the same as before, then he should have a higher ability, or in other words, he has a more special structure than before. After all, through generalization the theory becomes more special.

Majime  Mm, generalization specializes a theory. This implies that when you specialize a theory it might become more general, doesn't it? How can that be?

Morimori  Hahaha, that is the same as, "the one calling the other silly is sillier than the other".

Majime  Do you mean "the pot calling the kettle black"?

[Shinzuki looks at Majime and Morimori]

Shinzuki  Shall we apply this general theory to a particular example, or shall I finish with the general theory following to the rules decided by the majority?

[Majime and Morimori look at each other]

Morimori  Are you telling us to stop here and loose the opportunity to quest for truth?

---

[13] To formulate the player's knowledge and inference ability, we need to reformulate them in epistemic logic. Then we can study how the player makes a decision in a game based on the given knowledge and reasoning ability. This belongs to the interdisciplinary area of logic (foundations of mathematics) and game theory (economics). The number of researchers is increasing slowly internationally. For an introductory paper on this interdisciplinary field, see Kaneko M (2002) Epistemic logics and their game theoretical applications: introduction. Economic Theory 19: 7-62.

Majime    Whatever might be expected we should quest for truth. This is the duty for those gifted with noble minds, and moreover this is our choice even where there is no other choice.
[Together they say]
Majime and Morimori    Beautiful!

## Scene 4  The reversal of particularity and generality

[Shinzuki, looking at Majime and Morimori with a grin]
Shinzuki    Let me continue for a little more about generality of a theory as well as of a theorem.
Majime    Again, you are thinking to take the story in another direction, aren't you? Your face says that you are plotting something bad, Sir. Please keep it short.
Shinzuki    Okay! First, recall that when the axiomatic system $A = (A_1,...,A_n)$ is an extension of $B = (B_1,...,B_m)$, any statement that is provable in $B$ can also be proved in $A$. In this sense, the most extended axiomatic system is a complete theory. In such an axiomatic system, any statement or its negation is provable. The best example is an axiomatic system allowing essentially just one model. In this case, the axiomatic system is said to be categorical. When $A$ is an extension of $B$, $A$ has a smaller class of models than $B$. This may be interpreted as the contents of $A$ being more limited than those of $B$ [14].
On the other hand, when Theorem $C$ is a generalization of Theorem $D$, the models covered by Theorem $C$ are wider than those covered by Theorem $D$. Or, the contents of Theorem $C$ become richer and less limited than those of Theorem $D$.
Morimori    The argument sounds nice.

---

[14] When an axiom is added, the models for the axiomatic system decrease. In this sense, the extended axiomatic system $A$ restricts its contents compared to axiomatic system $B$. See Footnote 10.

## Scene 4 The reversal of particularity and generality

Shinzuki　Okay, now let me ask the following question, Morimori. Suppose that your head is tightly filled with a lot of thoughts. Should we compare your head with an extended axiomatic system or with a generalized theorem?

Morimori　Mm... when my head is tightly filled with a lot of things, what should that be? Maybe, putting a belt around my head to prevent it from an explosion, is not an answer, I suppose?

[Majime is disgusted]

Majime　Morimori, the only thing you have learned from Professor Shinzuki is how to pretend to be ignorant. And even that ignorance is borrowed from somewhere else. What an excellent student you are! So much nonsense! Now I will continue our discussions by myself.

Suppose, only supposing though, that your head is tightly filled with thoughts, Morimori. In such a case, compared to an empty state, you would be able to make more statements. Therefore, if I were asked to compare it with an extended axiomatic system or with a generalized theorem, I would certainly compare it with an extended axiomatic system.

In other words, in your head there exists an axiomatic system, and statements logically derived from the axiomatic system are coming out of your mouth.

Morimori　I see, my head contains an axiomatic system. All I need to do afterwards is to polish the logical ability to derive as many statements as possible from that axiomatic system in my head.

[Shinzuki is quiet for a while]

Shinzuki　The situation is slightly more complicated than that, Morimori. Often it is also necessary to extend the language or to add new axioms to the axiomatic system in your head. Since that is the goal of education and because you are a graduate student, you are still increasing your vocabulary and adding new axioms.

However, the direction I was aiming at with this discussion is not education, but it is to ask the question of which person should consider things more: one who was a less extended axiomatic system or one who has a more extended axiomatic system.

A hint for the answer to this question is to recall that a less extended system allows more possibilities, i.e., a bigger class of models. Applying this hint, we reach the following conclusion: An empty-headed person has to consider many more possibilities to reach one conclusion than a person with a more tightly filled head.

Morimori  I'm surprised this turned into the same story again. An empty-headed person thinks more and someone with a full head thinks less.

My purpose of coming to graduate school was to become a person who would think more to understand things better. After all it is better to keep my head empty in order to think more.

Majime  Morimori, you shouldn't be impressed by such a thing. I might be driven to insanity.

Sir, please return to the beginning of the story about the application of the general theory of the reversal of particularity and generality to a concrete example.

[Shinzuki is disappointed but becomes authoritative and sarcastic]

Shinzuki  Indeed, since we are economists and simultaneously game theorists, our social responsibility doesn't allow us to deviate too much from economics and game theory. So, let's consider the application to a concrete example in economics and game theory.

Let's consider an economic model with a time structure, and assume that the economic state is stationary over time. Suppose that we obtain a generalization by eliminating the stationarity assumption. You have often seen colleagues who are pleased with such a generalization.

Majime  Indeed, there are some economic theorists who are pleased with such generalizations.

Shinzuki  Now, the problem is as follows: An economic agent can easily perceive a past stationary state and predict a future stationary state, since they are the same. However, the elimination of the stationarity assumption makes the perception and prediction difficult. In this case, if we assume that the agent can predict or make a decision as in the original model, the generalization of eliminating the stationarity assumption implies a specialization of the agent's ability. The textbook understanding of "rational expectation" is an extreme example of this type.

Majime    Indeed, the elimination of the stationarity assumption may require a stronger ability of the agent.

Morimori    Could you please give us an example in game theory?

Shinzuki    An example in game theory? There must be an abundance of concrete examples in game theory. Let me think...

Okay, in game theory, the Folk Theorem for a repeated game may be a good example, or a bad example, of the reversal of particularity and generality[15]. The problem is to decide which, a good or bad example, it is.

Majime    Sir, please try to avoid any digression.

Shinzuki    Mm, I would like to discuss whether it is a good or a bad example...

The Folk Theorem states that Pareto optimal outcomes may be achieved in Nash equilibria when a one-shot game is infinitely repeated. When it is extended into an infinitely repeated game, the set of strategies becomes terribly large. The mathematical concept of a strategy in the repeated game is first to consider all the future possibilities and then to write down all the plans for the possible future events. The Folk Theorem needs the assumption that the players perceive this terribly large set of strategies. This theory itself is terrible, regardless of what aspects are considered.

Morimori    Bu... But the Folk Theorem has been considered very important in our field. Is it really so horribly terrible or terribly horrible?

---

[15] The word "Folk Theorem" is not mentioned in the literature of game theory before 1980, but was known and mentioned by certain game theorists already in the 1970's. For the Folk Theorem, see Osborne M, Rubinstein A (1994) A course in game theory. MIT Press, Cambridge.

Shinzuki    Mm, can I come up with a simple explanation? Okay, here is something I thought about before.

Usually when one discusses the Folk Theorem, one means an infinitely repeated game. But now consider the specific situation where the Prisoner's Dilemma is repeated 4 or 5 times. The one-shot Prisoner's Dilemma is described in Table 1.1. Alternatively, it is expressed as Fig. 1.4 in an extensive game[16].

Table 1.1: Prisoner's Dilemma $(g_1^1, g_2^1)$

| 1 \ 2 | $s_{21}$ | $s_{22}$ |
|---|---|---|
| $s_{11}$ | 5,5 | 1,6 |
| $s_{12}$ | 6,1 | 3,3 |

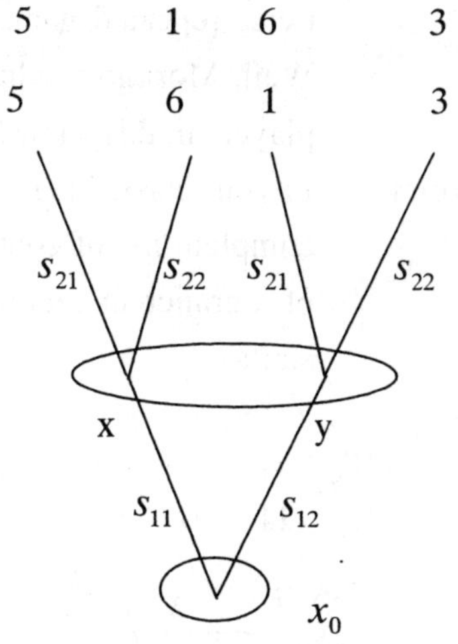

Fig. 1.4.

You understand how to read Table 1.1 and Fig.1.4. Each of players 1 and 2 has two strategies, and for example, when they choose $s_{11}$ and $s_{22}$ the payoff for each is 1 and 6 respectively. The pillows in Fig.1.4 are called the information sets for these players. The lower information set consists of only the root $x_0$. The upper one, a larger pillow, indicates that player 2 has obtained no information about the choice $s_{11}$ or $s_{12}$ of player 1 at $x_0$.

Morimori    Professor, this is elementary.

---

[16] The "Prisoner's Dilemma" will be explained in Act 2, Scene 2.

Shinzuki   Yes, it is. Now, consider the case where the game is played twice. We assume that after the first round, each player observes what both players chose. Fig.1.5 describes this twice-repeated game as an extensive form. The payoffs for the entire game are the sum of the payoffs of each round. For example, if $s_{11}$ and $s_{22}$ are played in the first round and then $s_{12}$ and $s_{22}$ are played in the second, the players receive payoffs (1,6) and (3,3) in these two rounds. Thus, the sum is (4,9). In this twice-repeated game both players have five information sets. Well, Morimori, what is the total number of strategies for both players in this game?

Morimori   In your class, I learned that a strategy in an extensive game is a complete list of contingent actions. This means that the choice of a branch at each information set is planned before the game starts.

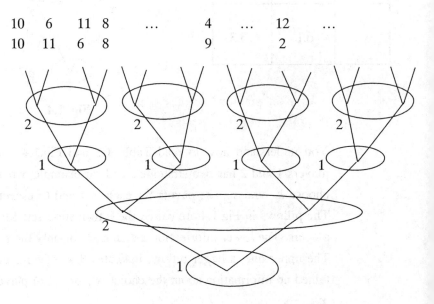

Fig. 1.5.

## Act 1 The reversal of particularity and generality in economics 33

In the game of Fig.1.5 each player has five information sets and has to choose one branch out of two at each information set. This choice is made independently. Thus, the number of strategies must be $2^5$. Is that correct?

Majime    Yes, it is, of course. Professor Shinzuki wants to consider the four or five times repeated Prisoner's Dilemma. The extensive game is hard to draw on the blackboard, but it is not hard to count the total number of strategies. First we count the number of information sets of each player, and the number of strategies is obtained by taking 2 to the power of that number. In the case of Fig.1.5, the number becomes $2^5$.

Morimori    In the three-times repeated Prisoner's Dilemma, each uppermost leaf of Fig.1.5 becomes the information set for player 1, and now it has two branches. Then the information set for player 2 appears and each has another two branches.

Majime    Morimori, you should think as if we engraft Fig.1.4 to the 16 leaves of Fig.1.5.

Morimori    I see, it is an engrafting. Now I understand. Mr. Majime, you are smart. To obtain the total number of information sets for each player in the three-times repeated game, we add 16 new ones to the previous 5. Now, the total number is 21. Therefore, the total number of strategies for each player is $2^{21}$ in the three-times repeated Prisoner's Dilemma. That is quite a big number.

Majime    I fear something strange may occur. But if I quit this calculation because of a fear of a bad result, I would lose the opportunity to be praised for the quest for truth. Let me continue a bit further.

In the game where the Prisoner's Dilemma is repeated four times, Fig.1.5 is engrafted into each leaf of Fig.1.5. Therefore, each of the 16 leaves has five information sets and the total number of new information sets is $16 \times 5 = 80$. By adding the

|  | 5 sets of Fig.1.5, each player has 85 information sets in total. Thus the total number of strategies for each player is $2^{85}$. Wait a moment. Let me try and calculate Avogadro's number to compare with this number. Mm, Avogadro's number is about $2^{79}$. Mm... what should I do now? |
|---|---|
| Morimori | By repeating the Prisoner's Dilemma only four times, each player has $2^{85}$ strategies. This number is a lot bigger than the Japanese national budget, and even bigger than Avogadro's number. How does each player think about such a big set of strategies? |
| Shinzuki | Okay, finally, in the five times repeated Prisoner's Dilemma, the number of information sets for each player is added up to 341 if I'm not mistaken. Then the total number of strategies is $2^{341}$.

As a mystery of physics, there seems to be a large number that prevails over the universe. That large number is said to be $10^{40}$. The number of the protons and neutrons existing in the whole universe is about $10^{40} \times 10^{40} = 10^{80}$ [17]. If we take 2 as its base, I think we obtain about $2^{266}$. However, this is much smaller than $2^{341}$, which we obtained for the number of strategies[18]. |
| Morimori | Hahaha, you got such a terribly large number by repeating the Prisoner's Dilemma only five times. Mm... an infinitely repeated game could be a terribly horrible or horribly terrible thing. Now I understand your intention, Professor.

Wait, I shouldn't be laughing. Is my generalization theorem okay? |
| Shinzuki | Even if we plug this large number to the budget constraint of the St Petersburg game, the expected payoff is about 342 |

---

[17] Cf. Davis PCW (1982) The accidental universe. Cambridge University Press, Cambridge.

[18] Some attempts have been made to avoid such a terrible feature. See von Stengel B, van den Elzen A, Talman D (2002) Computing normal form perfect equilibria for extensive two-person games. Econometrica 70: 693-715.

cents. This isn't much different from the conclusion of this morning when we employed the Japanese national budget.

What we learn from this, ultimately, is that we have to pay very careful attention to the relation between finite and infinite[19].

Well, let's consider a different example.

[Majime interrupts Shinzuki]

Majime     No, Sir, that won't be needed. I understood. I think I can construct other examples by myself. However, if we continue in this way, we would reach the conclusion that many of the latest developments in theoretical economics and game theory are devastated by the reversal of particularity and generality. Such an argument is applied to the theory of perfect equilibrium that was popular as a "refinement" in the 1980's, and we would conclude that it is an example of a reversal theory[20].

If we accept your opinion faithfully, we, theorists, can't write papers anymore. Okay, I should ask you, sir, about the reverse question: what general theory is free from your reversal of particularity and generality? What kind of concrete examples do you have in mind for that?

Shinzuki     Mm ... for that we shouldn't take easy jumps to new theories, and shouldn't pursue popularity. Instead, we have to make clear the basic concepts in economics and game theory while considering concrete examples for them. We have to give careful looks at the backgrounds of classical and orthodox problems, and have to look for their foundations. Based on these considerations, we should do research on how to develop new possibilities.

Majime     Sir, I didn't ask you such general principles.

You always say that "one should answer a question as clear and concrete as possible, and if one can't answer the question, one should say so". If I'm not mistaken, Sir, I asked you to provide

---

[19] A similar view is found in the literature of mathematics, for example, Kline M (1977) Why the professor can't teach, Chap.3. St Martin's press, New York.

[20] For refinements, see van Damme E (1991, $2^{nd}$ ed), Stability and perfection of Nash Equilibria, Springer, Berlin.

a concrete example of the general theory different from your particular examples.

[Shinzuki appears to be a bit guilty]

Shinzuki   I will do it some other time, for it is going to be a long story.

Morimori   Professor, you are beaten again, hahaha.

[Morimori starts to worry]

So after all, is my generalization theorem okay?

[Majime answers kindly]

Majime   The basic set that you generalized for your theorem is not directly, I say NOT directly, related to the perception of the economic agent in the model. Thus, even if we follow today's discussion, I think we can call your theorem a generalization. Sir, that is correct, isn't it?

Morimori, you should write a paper based on your generalization theorem. I would be happy to check and correct it. If it is well written, you could perhaps publish it in the *Journal of Theoretical Economics.*

Morimori   Thank you very much, Mr. Majime. Please help me correct my writing. So my paper will be published. That's great!

By the way, I have learned a lot in the discussions of yesterday and today. Having such discussions, you can obtain quite some results.

Shinzuki   In... indeed, this is the quest for truth, or more correctly the quest for universality via dialogues. But in fact, for you it is still too early. Plato says in his *The Republic* that one's twenties are a period to train basic physical strength and basic arts of thinking through physical education, music and mathematics. With that cultivated physical strength and arts one should take an active role in society in his thirties and forties. Based on those experiences one should pursue someplace higher and should quest for universality via the dialogue method in his fifties and later[21].

---

[21] Plato: The republic, book VII. Translated by Lee D (1955). Penguin Books, London.

## Act 1  The reversal of particularity and generality in economics

Thus, Morimori, you have to train your physical strength and get more arts, and you, Majime, you should be active in supporting society. And I could be starting dialogues.

Majime  But sir, unless you involve us in your dialogues, no economist older than fifty would join you. Mm... in addition, with only this kind of discussion it would be impossible to write research papers, and no productive people would remain in our academic profession.

Since I'm supposed to be an active member of our profession, I should produce research papers on what you, sir, call a special general theory or on a general special theory.

Morimori  My present job is first to train my body. The problem of gaining knowledge is whether to keep my head empty or to fill it up tightly...

[Shinzuki stretches his arms]

Shinzuki  Today was very productive. But it is about time to go home. What should I buy today? Mm, here is the list of my shopping written by my wife. Well, I will see you next week.

[Shinzuki leaves in a hurry]

[Morimori looks at the sunset through the window]

Morimori  Mr. Majime, today was very fruitful to me. I hope my research will be progressing in this way.

Majime  I hope so, too. To make progress and to write papers, however, we need to discuss things more concretely rather than always so abstract as today.

Morimori  I see, Mr. Majime. You can think in that way because you are ahead of me. But thanks to the fact that you are here, we can continue our discussion. Also because you give explanations to my basic questions, I could follow your discussions with Professor Shinzuki. So I will cooperate to make our discussions productive and fruitful.

[Morimori lifts a strong fist]

I will do my best!

## Scene 4  The reversal of particularity and generality

Narrator: This was a long story, and I'm already tired. The reader must feel tired too. After all, Mr. Shinzuki didn't tell us his own opinion on the best way to consider things. However, if we question him any more than this, I think we will get to listen to the long story based on Shinzuki's great research plans and huge amounts of labor. Philosopher Alfred North Whitehead wrote,

*"It is the characteristic of a science in its earlier stages ... to be both ambitiously profound in its aims and trivial in its handling of details"*[22].

This might imply that Shinzuki's story could be profound and trivial. Let's postpone what follows until you are ready for the long and trivial story.

Now, I must also hurry to do my shopping. It would be nice to have a partner like Shinzuki to help me do daily chores.

---

[22] Whitehead AN (1917) The organisation of thought, Chapter VI. Williams and Norgate, London.

## Act 2   Konnyaku Mondo and Game Theory

Narrator: The same characters of Act 1, Shinzuki, Majime and Morimori, discuss the relation between the Konnyaku Mondo and game theory. As far as I know, the Konnyaku Mondo is a Japanese comic story, and is meant to be an unclear dialogue. It is not at all likely for them to argue that game theory could be a konnyaku mondo. But if they go in such a direction, please reader, argue that game theory is not so. Still, I don't know exactly what the Konnyaku Mondo is about.

## Scene 1   Konnyaku Mondo

[Shinzuki, Majime and Morimori are having coffee]

Morimori  Professor, you often say, "Game theory is related to foundational problems of social science and therefore you have to think about the foundations of game theory". I also think the foundations are important, but what should we consider concretely?

Shinzuki  As usual, you ask such a difficult question. Okay, let me try to explain it to you. A while ago in Kyoto there was a workshop on epistemic logic and game theory, and both Majime and I participated in it. The epistemological side of game theory was addressed as one of the topics.

Majime  Yes, yes, I think the workshop was quite useful. I enjoyed especially the reception at the Kansetsu Hashimoto Memorial Hall[1]. Of course, the free discussion during the first day was useful, too. There the Konnyaku Mondo came up. Sir, what do you think about that discussion on the Konnyaku Mondo?

---

[1] Kansetsu Hashimoto was a central figure in the Kyoto school of traditional Japanese painting in the beginning of 20th century, and lived from 1883 to 1944. His studio, located in the old part of Kyoto, has been changed into a memorial hall.

Shinzuki   Ah... the reception was very nice: Viewing the Japanese garden, we had sake with good food. That sake was impressively delicious, wasn't it? Do you have any idea where it came from?

Morimori   Professor, you are making us talk about drinking. Mr. Majime, you too, please don't direct Professor Shinzuki to his favorite subject.

[Morimori turns towards Shinzuki]

Mr. Majime asked what you think about the Konnyaku Mondo. Also, could you please explain the meaning of the word "epistemic"?

Shinzuki   Indeed, we should be a bit more serious. First of all, the word "epistemic" means "about knowledge", but we use it in a wider sense meaning anything related to knowledge. You will understand the meaning better when you hear more and start using it. By the way, Morimori, do you know what the Konnyaku Mondo is?

Morimori   Is it a funny and unclear dialogue?

Shinzuki   Mm, it is often used to mean a clammy and slimy dialogue like a devil's tongue jelly.

Morimori   Wh... What is a devil's tongue jelly?

Majime   It is the English word for "Konnyaku".

## Act 2 Konnyaku Mondo and Game Theory

Shinzuki    Yes, it is dark brown, clammy, slimy and it looks like a devil's tongue, doesn't it?

Morimori    It is quite a big tongue. Mm... my girlfriend likes it because it is low in calories and high in fiber.

Majime    Morimori, stay focused.

Shinzuki    In fact, the Konnyaku Mondo is a Japanese traditional comic story in rakugo style. It is believed that the rakugo-teller, Hayashiya Shozo II, wrote it. It is not simply a clammy slimy dialogue, but it also suggests that our belief or knowledge is subject to subjectivity and falsity. In particular, it points out some of the difficulties in mutual understanding when two or more people are talking.

Majime    In the workshop we discussed how people believe certain things to be common knowledge among them. By "common knowledge" we mean that everybody knows the information in question, and in addition everybody knows that everybody knows it, and furthermore, everybody knows that everybody knows that everybody knows it, and so on. This repetition goes to infinity[2].

At the end of the workshop, we concluded that something becomes common knowledge among the participants but it might be the case that each participant attaches a totally different interpretation to that common knowledge.

Morimori    What do you mean by saying, "each participant attaches a totally different interpretation to common knowledge"?

Majime    In the free discussion, the participants all got the same information, which was common knowledge. However, each participant had a different understanding of what was discussed. This is what I meant.

Morimori    Thank you, I think I understand what you said.

---

[2] Research on common knowledge as well as game theory from the viewpoint of epistemic logic is now actively conducted internationally. An introduction to this interdisciplinary field is given in the paper: Kaneko M (2002) Epistemic logics and their game theoretical applications: introduction. Economic Theory 19: 7-62.

Majime   This is exactly what the Konnyaku Mondo implies. In addition to the conclusion we reached, the following conversations were exchanged and were even more interesting to me: The moderator, Mr. K, was ending the free discussion with the closing sentences:

*"The participants come from various disciplines such as economics, game theory, philosophy, mathematics, logic etc. Before the workshop I was anxious of where the discussion would go. It was beyond all my expectations that the discussion heated up so well. We discussed various different topics and heard different opinions from different disciplines. Nevertheless, the participants have a lot more in common now."*

Then one of the participants commented:

*"We express ourselves differently because we come from different disciplines. However, I think our goal of research is more or less the same, that is, to facilitate a better understanding of the world so as to improve our lives. After all, we think similarly. Thus, through the free discussion, we mutually understood each other. Today, I have ascertained my belief that if we spend enough efforts to communicate to other people, then we can reach mutual understanding even though we are from different disciplines."*

Morimori   That is a nice comment.

Majime   Then, however, one famous logician, Professor O, was there and made fun of this comment by saying,

*"You only think you understood each other, but, actually, each of us may be thinking about totally different things. So, we might have a konnyaku mondo".*

|  |  |
|---|---|
|  | This comment closed the free discussion leaving us with the feeling that the heated discussion was maybe nothing more than a konnyaku mondo. |
| Morimori | I see, one person thinks they understand each other, but another person doubts it. |
| Majime | No, no, you missed the point. All the participants believe that they had a meaningful discussion, while in reality each person is thinking of a totally different thing. That is what the Konnyaku Mondo tells us. |
| Morimori | Mr. Majime, if you know the story of the Konnyaku Mondo, could you please tell me about it? |
| Majime | Yes, but I'm not sure that I can do it well. After the workshop I wanted to read the original text and looked for it in the library. There was a book in which the Konnyaku Mondo was recorded from an oral presentation around 1890, and that is what I read. But it was written in an old style of Japanese, and it was really hard for me to understand[3]. Let me try to remember what I read. It is something like this: |

*"There was a temple where no monks were living any longer. A devil's tongue jelly maker, named Rokubei, lived next door. He moved into the temple and started pretending to be a monk. One day, a traveling Zen Buddhist monk passed by and challenged Rokubei to a debate on Buddhism. Rokubei had no knowledge on Buddhism and was not able to have a debate. He tried to refuse, but he could not escape and finally agreed.*

*The Buddhist dialogue started but Rokubei didn't know how to perform and therefore, he kept silent. The Buddhist monk tried to communicate to Rokubei in many ways. After some time, Rokubei started responding with gestures to the body movements the monk made. The monk took this as a style of dialogue and tried to answer in gestures, too. They exchanged gestures, and*

---

[3] Enomoto S et al (eds) (1980) Collection of meiji-taishou rakugo Part III, pp 61-70 (in Japanese). Kodansha, Tokyo.

*after some time, the monk told Rokubei, "your thoughts are profound and mine are of no comparison. I am very sorry to have bothered you". After saying this, he left the temple.*

*Hachigoro, a neighbor of Rokubei, witnessed the whole debate, and followed the monk to ask what had happened. The monk answered, "I'm not trained enough in Buddhist thoughts to compete with that master. Please convey to him my earnest apology for having left so abruptly". Almost as quickly as the words had left his lips, he ran away.*

*Hachigoro returned to the temple and asked Rokubei if he knew anything about Buddhism thoughts. Rokubei answered, 'No, I have no idea about Buddhism, but the guy is, in fact, a beggar and he talked badly about my jelly products. That's why I gave him a lesson'."*

[Morimori is impressed]

Morimori   Rokubei and the monk had a Buddhist dialogue through gestures and Rokubei won. Hachigoro asked both the monk and Rokubei how Rokubei won, and he received totally different answers from both of them. Is it written in the Konnyaku Mondo how these people interpreted the gestures?

Majime   Yes, it is. The monk, of course, interpreted the gestures according to Buddhism, while Rokubei interpreted the same gestures as describing his jelly products.

Morimori   I see, the two of them thought that they were having a meaningful debate, but, in fact, they interpreted the gestures exchanged in totally different ways.

The last comment by Prof. O meant that the participants shared many common thoughts through the free discussion in the workshop, though they were actually thinking of totally different things. Mm, what a cynical view it is!

Shinzuki   Morimori, you are becoming cynical. I thought you were just positive and energetic.

We can see a konnyaku mondo quite often, even in daily conversations. We may find an even more interesting situation. For example, a person whom you are talking to doesn't understand what you are talking about. You notice it clearly, but he says, "I see, I see", and he doesn't stop talking. No matter how hard you try to explain, the person repeats, "I understand what you want to say". Well, he unconsciously refuses the fact that something is incomprehensible for him.

The person who gave the comment almost at the end of the free discussion is a good, or bad, example of this.

Majime    Sir, again you become sarcastic the very moment you start talking. Maybe, other people think that you are such a person.

Shinzuki    Hehehe, you are right. I must be careful!

Morimori    But, Professor, I don't see any relation between the Konnyaku Mondo and game theory.

Majime    I don't see it either. It was discussed as a problem of epistemic logic even in the workshop in Kyoto.

Shinzuki    Really? I think the relation is quite clear. Okay, let me explain. The Konnyaku Mondo may be connected to the extant game theory in two ways. First, we find some Konnyaku Mondo-like aspects in game theoretical phenomena. Second, we often see konnyaku mondos in our profession.

Morimori    I understand neither.

Majime    It sounds interesting that game theoretical phenomena may have Konnyaku Mondo-like aspects. But, it is improper to say that we often see konnyaku mondos in our profession. I suppose that the latter is essentially the same as the cynical comment given by Prof. O.

Shinzuki    Yes, that is true, but both have the same structure as that of the Konnyaku Mondo. However, the second one, which Majime calls improper, is easier than the proper one.

The other day I mentioned that different circles of game theory attach different interpretations to the same mathematical formalism[4]. Consider a situation where people from different circles meet at a conference. Formal presentations are given using

---

[4] Cf. Act 1, Scene 3.

game theoretical jargon, i.e., mathematical language. The participants appear to understand and communicate with each other. However, the interpretations attached to the mathematical formalism may be totally different for people from different circles.

Morimori  Professor, do you have any concrete examples? I would like to know whom you are talking about and from which group these people are.

Shinzuki  Mm... quite a few people and circles belong to this type. You know about the field called *mechanism design* or *implementation theory*, don't you? The main concern of this field is to design an economic mechanism so that a social outcome attained by the mechanism based on the preference relations reported by economic agents is Pareto optimal. In order to have Pareto optimal outcomes, researchers in this field usually pursue possible rules of a social choice mechanism in which each agent reports honestly his true preference relation.

Economic theorists related to welfare economics love this kind of problems, and they form a circle. However, in the traditional circle of game theory it is assumed that the rules of the game, including the true preference relations of the players, are common knowledge, though this assumption is implicit and is not included in any mathematical formulation.

When two theorists from those different circles talk to each other, the mathematical part is mutually intelligible but their interpretations are totally different. There, it may happen that only the formal mathematical part is discussed but their intended contents are totally forgotten.

Majime  I have noticed it, too. But for now let's consider only theorists who care about the original intention of the mathematical formalism. Economic theorists of the other type are useless.

Morimori  Mr. Majime, you are harsh.

Shinzuki  Okay, when the preference relation of each player is included in the common knowledge, it makes absolutely no difference whether the reported preference relation is true or false, since the true one is known to everybody anyway. The set-up of mechanism design itself is meaningless from this point of view.

Morimori  That is awful. What do theorists in mechanism design think about this?

Shinzuki  They think that it is nonsense to assume that the preference relation of each agent is common knowledge. To a certain extent, I agree with them.

On one hand, when traditional game theorists make the common knowledge assumption, they have games with a small number of players in mind. By a small number, I mean two or three. On the other hand, the starting point of mechanism design is an economy with a large number of people. Accordingly, researchers in this field think it is inadequate to assume the preference relations of the participants are common knowledge.

Nevertheless, both theories adopt the Nash equilibrium as their basic solution concept. In other words, they use the same mathematical concept but attach a totally different interpretation to it.

Morimori  So which is correct?

Shinzuki  Mm... that is a difficult question. Both have shortcomings.

Morimori  Can you be more specific?

Shinzuki  Yes, let me give one concrete example. We will discuss the Nash equilibrium some other day[5].

Mechanism design targets problems with many participants. Now, assume that the number of participants is 100, and that there are two social alternatives $A$ and $B$. In the absence of indifference, either $A$ or $B$ has to be chosen. Each agent reports which alternative he prefers. The reports from the 100 participants form a vector consisting of 100 components of $A$ or $B$. Then the central government uses an aggregation rule based on these reports to determine which is socially desirable.

---

[5] Various interpretations of the Nash equilibrium concept are discussed in Act 4.

The aggregation rule is regarded as a function choosing $A$ or $B$ for each vector consisting of 100 components of $A$ or $B$. This function $F$ is expressed as

$$F : \{A, B\}^{100} \to \{A, B\}.$$

Morimori, how many vectors are in the domain, $\{A, B\}^{100} = \{A, B\} \times \cdots \times \{A, B\}$, of this function?

Morimori  That is easy. Each player chooses independently $A$ or $B$, so the number is obtained by taking the 100th power of $2$. So, $2^{100}$ is the number of vectors in the domain of $F$.

Hey, we have, again, a number of that form. Avogadro's number is about $2^{79}$, so this is a lot bigger. If the number of people is 341, the number of vectors in the domain is about the total number of protons and neutrons in the whole universe, isn't it? Such a function is impossible to calculate[6].

Majime  Unless the rules of a game are more concrete and simpler, no players would make rational decisions. We can't assume rational behavior for a player in the game you mentioned, though it is standard in our profession to apply mechanically the Nash equilibrium to any game. Sir, once more you succeeded in leading us to a negative conclusion.

Shinzuki  Hahaha, ... Anyway, we would immediately meet such a horribly terrible structure if we formulate the problem in a simplistic manner. It's my point that certain fields with different motivations use the same mathematical formalism and that this turns into a strange discussion. The reason for this is that the background and scope of each theory are not well examined. As long as these parts remain unclear, the discussion will often turn into a konnyaku mondo.

Majime  I think I understand what you are trying to say. We should move on to the other way that game theoretical phenomena might involve Konnyaku Mondo-like aspects.

---

[6] Cf. Act 1, Scene 4.

Shinzuki   Okay, ah, but it is already 3:30! I have one of these administrative meetings. I'm sorry but I have to leave.

[Shinzuki runs out of the laboratory]

## Scene 2   The Prisoner's Dilemma and the Battle of the Sexes

[Shinzuki comes back to the laboratory where Majime and Morimori are waiting]

Majime   Sir, it must be terrible to attend so many meetings.

Shinzuki   Yes, it is. I switch myself into energy saving mode and try to not think of anything except what is really needed. Some people are fascinated with formality. For such people, the superficial part is important, and its contents are secondary. The discussion with those people is nothing more than a konnyaku mondo. If everything were like this, my brain would start regressing. I need to talk about something more academic.

Where did we reach with the previous discussion?

Majime   Before going to the meeting, you said that you could connect the Konnyaku Mondo with game theory in two ways. The first one is that different circles of game theory may attach different interpretations to the same mathematical formalism. But people from different circles talk as if they understand each other. You gave an example of this. I think it is the same as the cynical view stated by Prof. O at the workshop in Kyoto.

The second one we were told is that there are Konnyaku Mondo-like aspects in game theoretical phenomena. You were about to start explaining this, but it was time for your meeting.

Shinzuki   I see. I have to start with Konnyaku Mondo-like aspects of game theoretical phenomena. Well, let's take the Prisoner's Dilemma and the Battle of the Sexes as examples. I will write the tables of these games on the blackboard. There are many textbooks discussing what are expected in those games[7]. Morimori, you know about these games, don't you?

Table 2.1: Prisoner's Dilemma $g^1$

| 1╲2 | $s_{21}$ | $s_{21}$ |
|---|---|---|
| $s_{11}$ | 5,5 | 1,6 |
| $s_{12}$ | 6,1 | 3,3 |

Table 2.2: the Battle of the Sexes $g^2$

| 1╲2 | $s_{21}$ | $s_{21}$ |
|---|---|---|
| $s_{11}$ | 2,1 | 0,0 |
| $s_{12}$ | 0,0 | 1,2 |

Morimori   Of course, yes, I do. The Prisoner's Dilemma and the Battle of the Sexes are elementary.

[Shinzuki speaks in an authoritative manner]

Shinzuki   Well, as an educator, I would be embarrassed if a graduate student under my supervision can't properly explain such elementary games. Of course, I believe that you, Morimori, can explain these games well, but as a small test, I ought to ask you to explain them.

[Morimori mimics Shinzuki's authoritative speaking]

Morimori   Well, as a student being educated by you, I would pass your test if I explain them with emphases on the parts that the educator always emphasizes. Of course, I believe, Professor, you always emphasize the structure but not the details. I ought to ask you to listen to my brief explanation.

Majime   Are you a parrot? Don't speak like that, just start explaining!

Morimori   Okay. In the Prisoner's Dilemma, both players 1 and 2 have two strategies each. That is, player 1 chooses either $s_{11}$ or $s_{12}$

---

[7] For these games, see Luce RD, Raiffa H (1957) Games and decisions. Dover Publications Inc, New York. For the historical details of the Prisoner's Dilemma, see Pundstone W (1992) Prisoner's dilemma. Bantoam Doubeday Dell Pub, New York.

in Table 2.1, and similarly player 2 chooses $s_{21}$ or $s_{22}$. If players 1 and 2 choose strategies $s_{12}$ and $s_{21}$ respectively, then their payoffs are $(6,1)$. The other cases are similar. Further, we assume that each makes his decision independently of the other. That is all about the rules of the game.

Then, the study of how each player makes a decision in the game is called a *solution theory* or a *decision making theory*. For the game of Table 2.1, regardless of whatever the opponent chooses, each player gets a higher payoff by the $2^{nd}$ strategy. In this sense, the $2^{nd}$ strategy is called the *dominant strategy*. The behavioral principle stating that you should choose a dominant strategy is called the *dominant strategy decision criterion*. If either player follows this decision criterion, then they choose $s_{12}$ and $s_{22}$, respectively, and then their resulting payoffs are $(3,3)$.

Shinzuki   Your explanation is simple and fine. Okay, I see you understand those games well.

Morimori   Thank you very much. In fact, I can explain even the history of such games as well as some interesting implications, which, Professor, you may like.

Shinzuki   I would love to hear those.

Morimori   Okay, I will try. In the 1940s when the Prisoner's Dilemma was introduced, or later in the 1950s, it meant a dilemma for each prisoner in his decision making.

The dominant strategy criterion recommends that player $i$ should choose $s_{i2}$ because it gives a higher payoff whatever the opponent chooses. In this case, the resulting payoffs are $(3,3)$.

Now, consider the last moment of player 1's decision making. We assume that player 2 is about to decide his $2^{nd}$ strategy $s_{22}$ as well as that he is expecting that player 1 would choose $s_{12}$. The situation is quite symmetric. In this case, player 1 starts thinking:

*"If I keep myself patient and decide to choose the $1^{st}$ strategy $s_{11}$, then player 2 might think the same, since the situation is symmetric and the resulting outcome would be (5,5), better than (3,3). Thus, I'm happy to choose $s_{11}$, and player 2 must be happy to choose $s_{21}$.*

*Hey, wait a moment, if player 2 chooses the $1^{st}$ strategy, then I should choose the $2^{nd}$ strategy, since the resulting payoffs will be (6,1). But if player 2 thinks the same, the payoffs would be (3,3) after all. So, should I be satisfied with the $1^{st}$ strategy and take (5,5)? But wait, should I just change my choice to the $2^{nd}$ strategy at the very, very last moment? Bu... but player 2 will think the same way..."*

Shinzuki  Wow, great! Definitely, you are not a reincarnation of a parrot. Continue.

Morimori  Lately, however, "the Prisoner's Dilemma" has changed to "the Prisoners' Dilemma".

Majime  You mean "a dilemma for the two players" rather than "a dilemma for each player", don't you?

Morimori  Yes, I do. In other words, it is not a dilemma in decision making for each prisoner, but instead it points out the fact that the final outcome (3,3) is worse for both players than the other pair (5,5). In economic terms, the resulting outcome is not Pareto optimal. The choice of the second strategy would be rational for the individual player in the sense that each player maximizes his payoff whatever the other player chooses. However, this doesn't lead to a socially optimal outcome.

Majime  That's correct. Global warming, car pollution in cities as well as many other environmental problems have the same or similar structures. In global warming, the influence of each individual on the global environment can be almost ignored, and thus everyone wants to use a car, air-conditioner, TV etc. for his convenience. However, since everybody acted this way, the earth

|           | has started to warm up and the welfare of the individual is declining. That is why "the Prisoner's Dilemma" is now called "the Social Dilemma". |
|-----------|---|
| Morimori  | I wanted to explain that part by myself, but Mr. Majime did it. Actually I have more to say.

The "Social Dilemma" may be okay as a name but it is strange to call it "the Prisoners' Dilemma". It is not so realistic to assume that the prisoners consider each other's welfare, since they must be selfish.

According to recent textbooks on game theory, each prisoner $i$ in the game of Table 2.1 will choose the dominant strategy $s_{i2}$, without having a dilemma at all. Though game theorists continue to call the game "the Prisoner's or Prisoners' Dilemma", no dilemma remains anymore. Even in global warming, nobody is in a dilemma. |
| Shinzuki  | Interesting, Morimori! It is a social problem that the social state or economic resource allocation is far apart from Pareto optimal outcomes. Nevertheless, nobody feels a dilemma. Even "the Prisoner's Dilemma" is not a dilemma anymore. Mm… what a cynical view! |
| Majime    | Morimori, you have certainly learned Professor Shinzuki's sarcasm. |

[Morimori sticks out his tongue and turns to the audience]

| Morimori | Both Professor Shinzuki and Mr. Majime praise me. |
|---|---|

[Morimori turns back to Shinzuki]

| Morimori | Next, let me also explain the Battle of the Sexes. |
|----------|---|
| Shinzuki | It is clear to me now that you understand these games well, so we can skip the Battle of the Sexes. |
| Morimori | Okay, Professor. |
| Majime   | But Sir, the Battle of the Sexes describes something directly related to our personal lives. Also I find that it has more in common with the Konnyaku Mondo than the Prisoner's Dilemma. |
| Shinzuki | That is right, I forgot about that. I was supposed to explain Konnyaku Mondo-like aspects in game theoretical phenomena. |

Morimori  So, please discuss them in connection with the Battle of the Sexes. In fact, I'm curious about what Mr. Majime means by "our personal lives".

Majime  Okay, let me explain the Battle of the Sexes in connection with our personal lives.

In Table 2.2, since the upper left (2,1) and the lower right (1,2) are Nash equilibria, they are suitable candidates for the final decisions of the players. It would be fine if either combination were successfully chosen. However, when we assume that each player is an independent decision maker, their choices may not lead to (2,1) or (1,2), that is, they may make a double cross. The resulting combination may be the lower left (0,0) or the upper right (0,0), though, if possible, each player wishes to achieve a more preferable combination.

A man and a woman are reciprocally attracted to each other, and so they would like to continue seeing each other. As usual, each person's consideration of the other differs slightly, which may lead to a failure of compromising. This is expressed in the lower left (0,0) or the upper right (0,0) in Table 2.2. If either

|          | would give way, it would be fine for them to achieve (2,1) or (1,2), but it is hard for them to change their own ways. Their considerations are mutually incompatible, and finally they separate. |
|----------|---|
| Morimori | No such thing is written in Table 2.2! |
| Shinzuki | Well, Majime is talking about his personal life. |
| Morimori | Are relationships between men and women so difficult? |
| Majime   | Sooner or later you will understand those complications and difficulties. |
| Morimori | Have you had similar experiences, Professor? |
| Shinzuki | Hahaha, I have experienced nothing of the kind. But now since the topic of problems between men and women has come up, let's consider it in connection with the Konnyaku Mondo.

According to the textbooks, players 1 and 2 in the Battle of the Sexes are a man and a woman, and the $1^{st}$ and $2^{nd}$ strategies are either to go to a boxing match or to go and see a romantic movie. The man wants to see boxing together with the woman, but if this were not possible, he would like to go to a movie with her. Not meeting the woman is painful for him. As for the woman, she wants to meet the man but if she can meet him either at the boxing match or at the movie, she would prefer the movie. |
| Morimori | That interpretation is written in many textbooks. The point is the possibility that they might fail to meet each other, right? But can't they avoid that possibility by calling each other on their cell phones? |
| Shinzuki | Yes, if they can use cell phones. However, here the problem is set up so that both the man and the woman make independent decisions. Of course, if it was only a problem of dating, they could talk it over on the phones. However, there are a lot of situations in which each person has to make an independent decision, and even if coordination is possible, at the last moment of choice, an individual being is to make an independent deci- |

| | |
|---|---|
| | sion. Since we consider such a problem now, we purposely talk about independent decision making. |
| Majime | Morimori seems not to have the slightest idea about problems between men and women, yet he has a girlfriend. I better continue here. |
| | In the situation stated earlier, the man has to make a choice either to go to the boxing match expecting that the woman will concede, or to go to the movie expecting that the woman will expect that he will concede. The woman is facing the same problem. |
| | Morimori, you said that they should discuss it over the cell phones but even when talking on the phone you don't always understand immediately what your partner wants, and neither the man nor the woman can say straightforward to their partner what he or she wants. That is the problem between men and women. Therefore, considering independent decision making has quite some meaning. |
| Morimori | Mr. Majime, are you telling me to study how men and women interpret the gestures of one another like in the Tale of Genji[8], as we did at senior high? Why don't you say clearly what you want to say? Interpreting gestures, I think, is an anachronism. |
| Shinzuki | Hahaha, it will be your problem soon. Now, I will explain a Konnyaku Mondo-like solution for the Battle of the Sexes. Please listen. |
| | Suppose that the man is considering that his payoffs are given exactly as in Table 2.2. He believes that this payoff table is common knowledge between him and the woman. In addition, he is conservative and believes that women should concede to men. He thinks this is a social virtue for women and that this virtue is common knowledge. Because of this, the man chooses to go to the boxing match. |

---

[8] The Tale of Genji is regarded as one of the greatest Japanese classic novels and was written by the female writer Shikibu Murasaki in the 10th century. It describes subtle and delicate stories of love affairs of Genji with many women.

Morimori   Wh... what an old-fashioned man! Which century did he come from?

Table 2.3: $g^3$

| 1 \ 2 | $s_{21}$ | $s_{22}$ |
|---|---|---|
| $s_{11}$ | 1,1 | 0,0 |
| $s_{12}$ | 0,0 | 2,2 |

Shinzuki   Please withhold your judgments and listen to the rest.

Suppose that the woman believes that the payoffs are given in Table 2.3 instead of Table 2.2. The woman believes that she and the man share a refined delicacy to enjoy a romantic movie, and that he actually doesn't want to see something savage like boxing. She believes this is common knowledge for both. On the other hand, the woman is also conservative: She believes it is a virtue for a man to be physically strong and that it is a virtue for a woman to understand it. And she believes that these virtues are common knowledge for men and women. That is why the woman is patient and concedes to choose the boxing match in the end. And the two of them meet happily at the boxing hall.

What I emphasize here is that each is considering a different game, different behavioral standard and different social virtue. But the result is a happy ending.

Morimori   Such convenient misinterpretations are one possibility and quite a coincidence at that.

Majime   Yes, it is one possibility, but I don't find it inevitable. Mm, maybe that is how a relationship between a man and a woman works.

|  |  |
|---|---|
|  | However, let's suppose that these two have been dating for a while. Then the woman realizes that the man is only coming to enjoy boxing and she is disappointed. She tells him, "Men should be physically strong while having a refined delicacy". This makes him uncomfortable with her. In the end, the two break up. |
| Morimori | Again, the relationship is broken. But, I'm starting to understand that there are difficult problems in a relationship between a man and a woman. |
| Shinzuki | Mm... it may happen that they continue to hold the false common beliefs and to get along together. For both of you, I think, it is too early to talk about the Konnyaku Mondo of a man and a woman. |
| Morimori | Professor, in your story both the man and the woman are very old-fashioned. Also, the woman seems inconsistent.<br><br>But misunderstanding may happen also in the Prisoner's Dilemma. The prisoners both think that the other is trying to pull their leg. Actually neither is thinking about the other at all. So they don't cooperate. In this case, there is no dilemma. This solves the Prisoner's Dilemma, doesn't it? |
| Shinzuki | Well, we could accept your argument as correct. However, I would like to draw your attention to one thing. In the above discussion, the game structure is the object of knowledge, and so are behavioral standards and virtues. This fact is almost forgotten in the present game theory. We'll discuss this problem another day.<br><br>For now, let's have some tea and continue the discussion in 15 minutes. |

## Scene 3  Games with incomplete information

|  |  |
|---|---|
| Morimori | I'm starting to understand that a konnyaku mondo may be seen in two different ways in game theory. Both mean that game theorists or game players attach different interpretations to the |

|  |  |
|---|---|
|  | same mathematical expressions. Those problems have been discussed in game theory, right? |
| Shinzuki | No, they haven't. |
| Morimori | Is that true? I wonder why they are not discussed? How should they be discussed? |
| Shinzuki | Unfortunately, the present game theory cannot capture a konnyaku mondo. Although game theorists or game players share the same symbolic expressions, each of them interprets these expressions in a subjective manner, and often the interpretation by one person totally differs from that by another. Such a problem cannot be captured in the present game theory. |
| Majime | I don't doubt you are right, but why? |
| Morimori | I believe the conclusion, but I don't see the reason for it. Could you please explain it? |
| Shinzuki | Both believe my conclusion, but neither finds a reason to believe, hahaha. I'm glad to explain it, but where shall I start? |
| Morimori | In Mr. Majime's lectures on game theory, he discussed knowledge and information in the theory of games with incomplete information. I don't think knowledge and information were mentioned anywhere else. I want to know why we can't capture a konnyaku mondo by games with incomplete information. |
| Majime | Exactly, I would like to know as well. |
| Morimori | If it cannot be explained in the theory of a game with incomplete information, then a konnyaku mondo is not a problem of game theory. |
| Shinzuki | Morimori, your conclusion is too abrupt and too conservative. Indeed, as you say, in the present situation of economics or game theory, the theory of games with incomplete information seems to be only the general theory of knowledge and information in games. However, we ought not to follow this theory blindly. |
| Morimori | No, I would never say such a thing. But what else is possible? |
| Shinzuki | Don't rush to conclusions. Today you asked the question of what the foundational problems of game theory are. To answer this we should clarify foundational problems related to game |

|   |   |
|---|---|
|  | theory, and then look for new and better perspectives for research on socio-economic problems. This means we should dig beyond or beneath the foundations of game theory. |
| Morimori | I see. But for the moment, could you please return to games with incomplete information? Because it is more relevant for me to understand standard subjects. |
| Shinzuki | Okay. Perhaps that will help us find what we should avoid. First, let's review games with incomplete information. By the way, don't you think the title "games with incomplete information" sounds lavishly serious relative to its contents? |

[Majime gets irritated]

|   |   |
|---|---|
| Majime | Sir, now I would like to hear the contents of your discussion rather than your lavish comments. |
| Shinzuki | Yes, I should explain the contents. Games with incomplete information try to capture the incompleteness of knowledge, i.e., "not knowing", as the uncertainty of information. The uncertainty of information is formulated as the possibilities of how the information could be interpreted. In addition, a probability is attached to each possibility[9]. |
|  | In this theory, possibilities are expressed as parameters, for example $\{a,b,...,z\}$, and a probability distribution over this set is assumed. The case of complete information is formulated simply as the singleton set $\{a\}$. Nothing is given besides $\{a\}$ about the contents of this knowledge or information. |
| Morimori | I think that is correct. |

[Shinzuki is elevating his voices, firming his fists]

|   |   |
|---|---|
| Shinzuki | If a theory says nothing about the simplest case, that is, the case of complete information, it can say nothing in a more complicated case. |
| Morimori | Sir, please explain this more concretely. |

---

[9] Cf. Myerson R (1991) Game Theory. Harvard University Press, Boston.

| | |
|---|---|
| Shinzuki | Alright. In games with incomplete information we assume that a player knows his own payoff function but he only knows the probabilistic possibilities of his opponent's payoff function. |
| | Now, consider a situation where the Prisoner's Dilemma is played as an actual game. Each player $i = 1,2$ knows his own payoff function $g_i$, but doesn't know what his opponent knows. Suppose that player 1 has the partial knowledge that the payoff function for player 2 is $g_2$ or $h_2$, where $h_2$ is player 2's payoff function in the Battle of the Sexes. As I mentioned before, we take that "not knowing" is expressed by means of a probability. Suppose that the probability of $g_2$ is assumed to be $1/2$ and the probability of $h_2$ is $1/2$. |
| Majime | That is quite a standard explanation. |
| Shinzuki | No, it is faithful to the mathematical formulation, rather than standard. Now I will try to give a slightly more standard explanation. |
| | Player 1 knows that he has payoff function $g_1$ of the Prisoner's Dilemma. But he doesn't know if player 2 has payoff function $g_2$ of the Prisoner's Dilemma or the payoff function $h_2$ of the Battle of the Sexes. In addition, player 1 considers how player 2 will think. The player 2 in the mind of player 1 might have $h_2$ as his payoff function. Also, this imaginary player 2 doesn't know whether the payoff function for player 1 is $g_1$ or $h_1$. The symmetric argument applies for player 2. |
| | Player 1's incomplete information over $g_2$ or $h_2$ is expressed in terms of a probability distribution of $1/2$ for $g_2$ and $1/2$ for $h_2$. In the standard explanation, it is assumed that both players know this probability distribution, and moreover, that the probability distribution is common knowledge. |
| Majime | Certainly, your explanation has become more standard. |
| Shinzuki | By the way, Morimori, where is that common knowledge assumption written in the mathematical formulation of a game with incomplete information? |

| | |
|---|---|
| Morimori | I have seen that kind of sentence in many papers, but does any mathematical part correspond to it? |
| Shinzuki | In fact, the common knowledge assumption is not formulated mathematically. You can't regard it as a mathematical assumption, though it is the central part of the argument. It is an assumption that you should interpret it in that way, but nothing is said rigorously. |
| | This is regarded as a custom in the mainstream of our profession. It is quite similar to saying "It would be safe to cross the road even at a red signal if you go together with everybody else". |
| Majime | I thought that it was a bit suspicious. But can we now return to the discussion about the theory of games of incomplete information? In order to analyze such games, Harsanyi defined the Bayesian equilibrium[10]. Morimori, what does it look like in the game Professor Shinzuki described just before? |
| Shinzuki | It takes quite some time to calculate it. Let's give this as homework to Morimori. Morimori, please calculate the Bayesian equilibrium of the game, with the help of some textbook. |
| Morimori | Yes, I will do it later. In fact, I wanted to calculate it now. |
| Shinzuki | No, no, it is your homework, Morimori. Instead, I would like to consider how to handle complete information. What happens when we assume complete information in the game mentioned before? |
| Morimori | Mm, according to your explanation, each player knows his own payoff function anyway. For the opponent, complete information means that only one possibility remains. In other words, taking the example on the blackboard, it becomes simply the Prisoner's Dilemma. |
| Shinzuki | That is correct. |

---

[10] Harsanyi JC (1967/68) Games with incomplete information played by `Bayesian' players, Parts I,II, and III. Management Sciences 14: 159-182, 320-334, and 486-502.

## Act 2  Konnyaku Mondo and Game Theory

Morimori  I see. The case of complete information is the simplest case. So, you claimed, "If a theory says nothing about the simplest case, it can say nothing useful in a more complicated case."

Shinzuki  Exactly.

[Majime is challenging]

Majime  I think your argument is quite one-sided. Many other people have positive opinions on what you are criticizing or at least some justification for it. Please listen to the following argument in favor of games of incomplete information.

*"In principle it must be possible to have a complete description of the society or social state. However, the theory of games with incomplete information is constructed independently from such a description. The problem is what the number of possibilities for the social state is and how people have probabilistic assessments over those possibilities, rather than the contents, or internal structure, of each possible description. The theory requires only that the player should know the number of possibilities and the probability distribution over the possibilities. When necessary, it must be possible to have a more detailed description of each social state."*

Morimori  It is a good justification.

Majime  Then, this continues as follows:

*"In this sense, the theory of games with incomplete information must be independent from the detailed description of each possible social state. Thus, the theory of games with information is general enough to cover research based on more detailed descriptions. Also, in principle, more concrete problems involving knowledge or information can be handled as an application of this theory."*

Morimori  Ah... games with incomplete information are well justified. Then what do you think, Professor?

| | |
|---|---|
| Shinzuki | Mm... I have a slight headache. Such a justification is constructed with a great deal of thought. It is not easy to refute it. Nevertheless, it is nothing more than an excuse after all.<br><br>First, the problem is that the description of a social state is not so trivial as to be counted just as one possibility. Secondly, the relation between the description and its possible contents is not so simple as to be ignored. |
| Morimori | Please explain more concretely. |
| Shinzuki | Here, I think I should use an analogy. This is similar to saying, "I can do anything I want if only I would try". Maybe, some people say, |

> *"Axiomatic set theory has the generality of covering every field of extant mathematics. Thus, research in axiomatic set theory is sufficient for all other mathematical sciences including game theory."*

| | |
|---|---|
| | If this was the case, then the person who constructs such a justification should only do research on axiomatic set theory. However, I don't think that decent researchers of axiomatic set theory would say such a thing. |
| Morimori | Indeed, it is quite bad to say, "I can do anything I want if only I would try", while not trying.<br><br>When I was a child I didn't study so much as my bright sisters. My grades were not like theirs', and my mother often scolded me and told me, "Study more". Then my father always comforted me with the words, "Genki, if you try, you can do it, but you are just not trying". My eldest sister got angry with my father, saying, "But it is useless if he doesn't try". |
| Shinzuki | But, Morimori, lately you are studying quite a lot. |
| Majime | Yes, he is. But let's turn the conversation back to games with incomplete information. I can offer another justification to your previous argument in the case of complete information. |

*"The case of complete information is not of our interest. So we shouldn't be bothered with that case. When a case is expressed as a combination of games with complete information, it is a problem of truly incomplete information, which we are now interested in."*

Morimori  Hahaha, are you interested in a combination of things you have no interest in? That is strange. If you look closely at the structure of the theory of games with incomplete information, the simplest case is a game with complete information. Thus, we should be able to have a meaningful discussion about that simplest case.

Shinzuki  I agree with Morimori in his conclusion. It would be ideal to treat the case of complete information in a meaningful manner, and also to be able to discuss accurately the case of incomplete information.

However, your word "strange" is strange. For example, both sugar and alcohol are molecules composed of carbon, hydrogen and oxygen. The property of sugar being sweet is not that of either carbon, hydrogen or oxygen, but it is of sugar molecules. The atoms, as building elements, are not sweet, but the sugar molecules are sweet as the compounds of those atoms. The composition of an alcohol molecule differs slightly from that of a sugar molecule, and its property is not "sweet". In sum, you might be interested in a combination that is composed of things you are not interested in.

Incidentally, alcohol can be divided into methanol and ethanol. I have no particular interests in carbon, hydrogen, oxygen or even methanol, but somehow I feel a great temptation with ethanol. The sake we had at the reception in Kyoto is composed of water, ethanol and something impure and uninteresting, and it caught our attention.

Morimori  You are great, Professor. But again you are making us talk about drinking.

[Morimori thinks for a while]

But I'm able to think of other cases as well. For example, a bookstore selling only boring books is boring after all. We can compare this example with a case of incomplete information. Consider a lottery: With probability 1/2 you get a boring book of game theory, and with the other probability 1/2 you get a boring book of economics. This lottery must be boring anyway just like the boring bookstore.

Shinzuki    Certainly, Morimori, you are clever.
[Majime gets angry]
Majime    What is this? Are you having a contest to impress each other? You are just happy to divert the topic to something else.
[Majime sighs deeply and says to himself]
     Toru, calm down!
     Okay, let me summarize our discussion. The theory of games with incomplete information says nothing about information in the case of complete information. In the case of incomplete information, all the possibilities are known and even the probability distribution over those possibilities is known. This is too much for the description of "not knowing".

|  |  |
|---|---|
| | In short, "the case of complete information is purely trivial, and the case of incomplete information is outrageously bizarre". |
| Morimori | Mr. Majime, you are radical. |
| Majime | Even I can say a funny thing if I want. You know, "Still waters run deep". Mm... might I say too much? |
| Morimori | No, no, you are right. It is a good description. |
| Majime | Thanks a lot, Morimori. You are a nice person. |
| | Okay, let's accept the incapability of describing the interpretations or misunderstandings that appear in the Konnyaku Mondo with a form of incomplete information. |
| | Sir, what is the essential difference between knowledge handled in the theory of games with incomplete information and knowledge in the Konnyaku Mondo? |
| Shinzuki | The essential difference, you ask? |
| | Mm, in the Konnyaku Mondo, gestures are expressions, and Rokubei and the monk interpreted those gestures. The interpretation is totally different for each person. |
| | On the other hand, in the theory of games with incomplete information, information is represented as a set of possibilities, but its symbolic expression is not considered. Therefore, there is no distinction between the expression of information and the intended content attached to the information. Perhaps, this is the essential difference. |
| | In fact, mathematical logic teaches us that any symbolic expression has an infinite number of interpretations. The representation of information in terms of possibilities deviates a lot from this teaching, and comes from the naive belief in mathematics that everything is expressed as a set. |
| Majime | Your last comment sounds too strong to me, but I understand the others. In the theory of games with incomplete information, we can't make a distinction between the expression and its intended interpretation. This distinction is critical in the Konnyaku Mondo. We need mathematical logic to have this distinction, don't we? Is that the reason why we discussed the |

| | |
|---|---|
| | Konnyaku Mondo at the workshop "Epistemic Logic and Game Theory" in Kyoto? |
| Shinzuki | That is right. In mathematical logic, symbolic expressions and the intended meanings are totally separated, though separation raises many delicate problems. |
| Morimori | Even I'm starting to understand that the Konnyaku Mondo cannot be explained in the theory of games with incomplete information. |
| | But, Professor, the Konnyaku Mondo-like solution you gave for the Battle of the Sexes sounds still unrelated to game theory. Moreover, I don't feel that gestures or their interpretations are important to game theory. If this is your claim, Professor, could you please tell me the reason? |
| Shinzuki | You are raising another difficult question. Mm... shall I start with an historical explanation of the problem? Maybe we will find something while I'm talking. |

[Shinzuki looks up and says as if recalling something]

John C. Harsanyi introduced games with incomplete information in order to analyze a situation where the participants in a game do not completely know the structure of the game. In a parlor game, players know each other's payoffs completely, and there is no problem for the classical game theory. However, when addressing socio-economic problems, we cannot assume that the members of society know the social structure. Harsanyi introduced games with incomplete information so as to handle such situations.

Of course, one should also discuss how knowledge about the game structure is acquired. When epistemological foundations for such situations become clear, it becomes possible to study various socio-economic problems.

Harsanyi was awarded the Noble Prize for Economics in 1994, since his theory appeared to enable us to analyze such situations.

Morimori  By the way, Professor, it is almost 6 o'clock. Don't you have any shopping to do today?

[Shinzuki is surprised a bit]

Shinzuki    Wu... I almost forgot. Today my wife and children are not at home so I have no shopping to do. So we can continue the discussion. But I feel tired. Shall we have some coffee?

## Scene 4  Rashomon

[A cup of coffee energizes Shinzuki]

Shinzuki    Okay, let me continue a bit more with the historical review of game theory.

Unlike a parlor game, in a socio-economic situation, people have very limited knowledge about the social structure such as social rules, payoffs and even who the participants are. In such a situation, people realize what kind of behavior is suitable through experiences and by being part of the situation. They will come to understand vaguely how people would respond to each other, or how they might think etc. However, it would be almost impossible for a person to grasp the social structure exactly.

Morimori    I think that is correct.

Shinzuki    The theory of games with incomplete information was made in order to analyze such a situation. However, as already discussed, the theory suffers from the serious shortcoming that it doesn't distinguish between symbolic expressions and their intended contents. Moreover, the theory mentions nothing about the emergence of beliefs or knowledge about the game. Ideally, we could have the origin and emergence of beliefs and knowledge in the scope of the theory.

Morimori    Professor, what do you mean by the origin of beliefs or knowledge?

Shinzuki    Some beliefs are *a priori* granted. Some others have been formed from arbitrary reasons. Besides these, beliefs and knowledge are somehow based on direct individual experiences or what others have taught through personal relations in social environments or through systematic education. Thus, we have

|          | three categories of knowledge and beliefs: those *a priori* granted, those formed from arbitrary reasons, and those based on someone's experiences. To make our discussion simpler, we will consider only knowledge and beliefs of the third category. |
| Majime   | In the third category, some are directly experienced and some others are derived from direct experiences. Here, the person infers beliefs or knowledge by induction from direct experiences. |
| Shinzuki | Exactly. |
| Majime   | The word "induction" differs from "mathematical induction" and means that a general law is inferred from limited experiences, right? |
| Shinzuki | That's right. "Mathematical induction" has "induction" but, in fact, is a rule of deductive reasoning. In empirical sciences, induction is the key inference. |
| Majime   | This is also the original philosophical meaning of "induction". |
| Shinzuki | Yes, it is. Nevertheless, we are talking about ordinary people's inductive inferences, not about scientists'. The scientists' inductive inferences should follow rigorously certain statistical criteria. Ordinary people use induction much more boldly, that is, they often derive a general conclusion from a very small number of instances. For example, some people have the belief that every Japanese tourist has a camera, inferred from seeing only two Japanese tourists with a camera.

This is important, particularly in considering the formation of individual beliefs and knowledge about a social situation. People may infer some beliefs about the social structure from their very limited experiences. Here, since the social structure is far more complicated than what may be inferred from individual experiences, it is almost impossible to have correct beliefs. A lot of arbitrary guesses or judgments are added in order to reconstruct a belief on a complicated structure. |
| Majime   | Sir, then your third category is not mutually exclusive with the second category of beliefs formed from arbitrary reasons. |
| Shinzuki | You are right. They are not exclusive. Even the boundary of each category is unclear. |

|  | But it's my point that people who have been in the same place and have had the same experiences end up making totally different interpretations about the same experiences, since a lot of arbitrary elements may be involved. |
|---|---|
| Majime | Ah, I see. Here the Konnyaku Mondo comes into your story. |
| Shinzuki | Now you must understand the Konnyaku Mondo is related to game theory. |
| Majime | Yes, I think so. When we say the same experiences, that is only the surface and each person makes and attaches his own interpretation to the experiences. In terms of the Konnyaku Mondo, the sender of a gesture attaches an interpretation to the gesture, and the receiver shares this gesture but may interpret it in a totally different manner. |
| Morimori | Ah... it sounds like information exchange in extensive games, doesn't it? Finally I'm starting to understand the relation between game theory and the Konnyaku Mondo. Indeed, it would be nice if we could develop such a theory. So, what kind of concrete applications can we think of? |
| Shinzuki | Mm, the moderator at the workshop in Kyoto, Mr. K and his collaborators seem to be working on the theoretical part from the viewpoint of mathematical logic. |
|  | As a concrete application, one can think of the "self-fulfilling prophecy" as discussed by the sociologist Robert K. Merton. Merton gave, as examples, the great recession in the 1930's and racism in the US[11]. When someone starts talking about an expectation or a wish and shows a certain behavior, though no reality exists at that time, his talking and behavior work as a prediction to other people, and in the end the expectation or wish could be realized[12]. |

---

[11] Merton RK (1949) Social theory and social structure. The Free Press, London.
[12] For a game theoretical consideration of discrimination and prejudices, see Kaneko M, Matsui A (1991) Inductive game theory: discrimination and prejudices. Journal of Public Economic Theory 1: 101-137.

| | |
|---|---|
| Majime | Sir, I remember the movie, "Rashomon", by Akira Kurosawa[13], which suggests the opposite phenomenon. In the self-fulfilling prophecy, a verbal statement creates new reality, while in Rashomon, everybody describes the same reality in a different manner. This is more similar to the Konnyaku Mondo than the self-fulfilling prophecy, isn't it? |
| Shinzuki | Oh, yes, I think so. Akira Kurosawa was one of the few great movie directors in the $20^{th}$ century of Japan, and Rashomon is one of his masterpieces. I saw the movie several times. |
| Majime | May I tell the story? |
| Shinzuki | No, please let me do it, since I have been a great fan of Kurosawa as well as Rashomon. Toshiro Mifune and Machiko Kyo acted in the movie, and they were young, energetic and very sexy. |
| Morimori | When was the movie made? |
| Shinzuki | It was made and released in 1950, and was awarded the grand prix in the Venice International Film Festival in 1951. Anyway, the story goes as follows. |

[Shinzuki makes his voice wide and strong]

*"It was in the $9^{th}$ century of Japan that a certain incident occurred in the deep thick forest not very far from the capital city, Heian-kyo[14]. A wild thief called Tajomaru mugged a married couple, raped the beautiful wife called Masago, and then murdered the husband".*

Incidentally, muscular Toshiro Mifune was the wild thief, Tajomaru, and glamorous Machiko Kyo was the beautiful wife, Masago. The story continues as follows:

---

[13] The original story for Kurosawa's "Rashomon" is "In the grove" by Ryunosuke Akutagawa. It is included in: "Rashomon and other stories by Ryunosuke Akutagawa". Translated by Kojima T, Charles E (1952). Tuttle Co, Tokyo.

[14] "Heian-kyo" is the former name of Kyoto used before the Meiji restoration in 1868.

*"A villager was hiding behind trees and witnessed the whole incident. One day after, the thief was arrested, and he and the others were in court. They were asked about what happened in the forest. First, the thief and wife gave their accounts of the incident. Their stories differed in various points emphasizing what was important for each of them. Then the villager gave his account, and the story was, again, different. All stories sounded somehow twisted. Finally, the dead husband also recounted the incident through the mouth of an old medium. It also differed from the others, and simply presented another twist to the court. In the end, nobody in court believed anything about the incident."*

Morimori  Is it a violence movie or a court case drama?

Majime  No, Morimori, the emphasis of the movie is that the people who saw the same incident interpreted it in totally different ways.

Shinzuki  Exactly. Rashomon depicts a problem very similar to the Konnyaku Mondo.

Well, but Rashomon includes some other interesting aspects that are not found in the Konnyaku Mondo. Not only did the four people interpret in different ways what they saw, but they also altered what they saw. When the dead husband recounted the incident through the medium, his story was biased toward his wishful thinking, and also the story of each of the others suffered from a similar bias.

Morimori  I see. Each person altered what he or she saw in a manner that is comfortable. Does this alteration happen consciously?

Shinzuki  No, of course not, it must happen mainly unconsciously. It is an unconscious rationalization. Rationalization of one's own research with a difficult argument such as the one Majime gave is based on the same psychological nature. Before this kind of process the researcher doubted that his research would make sense, but after the justification process, he has succeeded in eliminating his doubt unconsciously.

|  | There is one thing curious to me in Rashomon. Normally, when you are dead, or when you are about to die, you have no desires anymore and you could be expected to speak the truth. But in Rashomon even the dead husband altered the story so as to keep his self-esteem. Is this true? |
|---|---|
| Morimori | I have no idea, since I have never died. |
| Shinzuki | You are right. It was a bad question, sorry. |

[Shinzuki thinks a little while]

|  | Mm... yes, yes, I wanted to point out the objectivity and subjectivity involved in the Konnyaku Mondo and Rashomon. In the Konnyaku Mondo, the gestures exchanged remain objective and the interpretations are subjective. In Rashomon, even this distinction becomes unclear. |
|---|---|
| Majime | I will paraphrase, in game theoretical terms, what you said. |

Consider a game with a few players where each player has no knowledge about the other players' payoff functions. In such a situation, it is adequate to assume that after the game is played once, the actions taken by the players become common knowledge. But since the payoff function of each player is unobservable by the others, it is not revealed and remains subjective.

If we take your Rashomon aspect into account, things may become more obscure, because the objective observations of actions taken may involve subjective elements.

Shinzuki   Exactly, now we have various important points. I should summarize them.

The action taken may become common knowledge, but the payoff function of each player or the reasoning process is a different story. With respect to these, the extant game theory simply starts with the assumption that there are utility functions, which each player knows. However, in fact, even though a player thinks of the payoff functions of the others, they may be in his own imagination.

In addition, Rashomon tells us that even the observed may involve subjectivity. We unconsciously alter the observed infor-

|  |  |
|---|---|
|  | mation for our own comfort. Thus, an erroneous interpretation occurs not only about others but also about ourselves. |
| Morimori | We can't trust even ourselves. Then what should we believe? |
| Shinzuki | Of course, we can trust neither ourselves not others. So at least, we should be conscious of asking whether or not we are inclined to make a comfortable interpretation for ourselves. |
| Majime | Sir, you are so agnostic. |
| Shinzuki | No, originally I was not agnostic at all. I was foreboding that I might turn into an agnostic, and indeed, as a result of my logical pursuit, I came to be an agnostic. |
| Majime | Isn't this similar to the Oedipus quest, which was mentioned before[15]? Foreboding that he cannot search for all the truth, he continues searching anyway, but eventually he falls into agnosticism. |

[Shinzuki is pleased]

| | |
|---|---|
| Shinzuki | Exactly. By the way, did you notice that a person who does an Oedipus quest is psychologically quite opposite to a Rashomon person? The former cannot stop his quest for truth, even knowing that it will result in a tragedy, while the latter looks for his benefits, adjusting unconsciously his own truth. Which is more respectable and more valuable? And which do you think is closer to the truth? |
| Morimori | I have no idea. But Professor, I feel as if you want to lead us to the conclusion that you belong to the more respectable and valuable one. |
| Shinzuki | Hahaha, I incline to have my own wishful thinking! I should be careful. |
| Majime | Sir, people interpret their own actions as well as others' in their own way, and these are full of mistakes and errors. But isn't it fine if society still works in that way? |
| Shinzuki | Mm, but there are a lot of problems. Mistaken and erroneous interpretations may seem simply comical at that point. However, these interpretations form the subjective part of each per- |

---

[15] Act 1, Scene 2.

son, and he will make decisions based on that subjective part. Behavior of those people affects society. The cause for the social result may be mistaken and erroneous interpretations.

Thus, what we have discussed is related to not only foundations of social science, but also to various actual socio-economic problems of today.

Morimori　When you say present-day socio-economic problems, do you mean the ones mentioned by Merton, for example, the self-fulfilling prophecy, and discrimination and prejudices?

Shinzuki　Yes. I do. And I think that other examples are juvenile delinquency such as bullying, drugs and motorcycle gangs, or middle-age idiocy such as negligence and irresponsibility of politicians, officials, scientists or doctors.

Majime　Well, Sir, you mentioned various socio-economic problems, but I don't understand well how they are related to the discussion. For example, how is "bullying" related to a mistaken and erroneous interpretation?

Shinzuki　Bullying is often directly related to an excuse or rationalization, and an excuse is based on a mistaken and erroneous interpretation.

First, "bullying" is accompanied by a structure of rationalization after the fact. For example, children or some adults often justify their behavior by saying, "the one being bullied gave us a reason to do so and therefore, he is responsible". They explain the "bullying" as a cause-effect event to justify themselves. They explain it as if it is an inevitable event for them. But they ignore the fact that their decision making entered the cause of the event. The responsibility should lie in their decision making. This is where a mistaken and erroneous interpretation enters the process, though I think much more erroneous elements are involved in their justifications.

I also think that responsibility lies with adults. It is because adults teach the attitude that everything is acceptable if there is an explanation for it in causal terms. Children learn this attitude and try to find excuses. But we should say, "bullying is wrong",

|         | as a categorical imperative. That is how moral principles should work. |
|---------|---|
| Majime | Okay, I'm starting to understand quite well what you want to say, Sir. We claim that such issues are related to the foundational problems of game theory, don't we? I never thought about those. |
| Shinzuki | That is why foundations of game theory are important. |
| Majime | Sir, you are quite a boaster. |
|         | Anyway, I'm starting to understand the reason why the Konnyaku Mondo was discussed at the workshop in Kyoto. I took it simply as a problem of epistemic logic. Even though I made a different interpretation of the discussion from your intention, I agreed with you that the workshop was interesting. As a matter of fact, we had a konnyaku mondo. |
| Morimori | Yes, yes, in my case, I almost always agreed with Professor Shinzuki that the foundation of game theory is important without thinking over the contents of it. So, we also had a konnyaku mondo. That is the same as Mr. Majime. |
| Shinzuki | Mm, you are right. But something differs slightly from the Konnyaku Mondo. In fact, I noticed from the beginning that both of you and I were discussing the problems with slightly different contents. That is why I said at the beginning of today's discussion, |

*"We may find an even more interesting situation. For example, a person whom you are talking to doesn't understand what you are talking about. You notice it clearly, but he says, 'I see, I see', and he doesn't stop talking. No matter how hard you try to explain, the person repeats, 'I understand what you want to say' ".*

[Majime and Morimori look at each other]

| Majime | Is this the conclusion you arrived at after our long discussion? |
|---|---|
| Morimori | Now I totally understand that it is common knowledge that you are truly sarcastic. |

[Shinzuki looks happy]

Shinzuki   Pff, today was very fruitful. Except for that meeting which made my head regress. Tonight I'm going out for dinner because my wife and children are not at home. Do you want to join me for some ethanol, uh sorry, sake? Maybe we may not find sake like the one we had at the reception in Kyoto, but some must be available. Moreover we can continue our discussion. It would be more enjoyable if you try to revenge yourself on me.

Narrator: This time, Shinzuki talked quite a lot about the contents of game theory. I was surprised to learn that foundational research is so ambitious. But it might sound just like an exaggeration. Incidentally, Majime and Shinzuki gave good descriptions of the classic stories. May I compete with them?

> Oh! Set us free
> O let the false dream fly,
> Where our sick souls do lie
> Tossing continually[16]!

Ah... this doesn't fit well here, does it? I should compose a poem myself, but this is difficult for me. Nevertheless, you, reader, understand I'm not just an anonymous narrator but do have an important role in these plays. Anyhow, I'm relieved to hear that game theory itself is not a konnyaku mondo. Now, the three are heading off to a bar. Please enjoy your sake and your discussions, but don't drink too much.

---

[16] Arnold M (1905) Poetic works of Matthew Arnold. Stagirus, p 40. McMillan and Co, London.

## Act 3  The market economy in a rage

[The poet appears silently before the curtain]
Man quested for a plateau and found a lost world
Dazzling lights, freezing winds and sharpening senses

Followers climbed and explored the plateau
Prevailing beauty and perpetuating truth

More and more came and exploited the new world
Freedom exhausted but everlasting longing

Man wished the greatest happiness of the greatest number
Wishes unanswered but all fascinated with games in dark caves
[The poet leaves the stage quietly]

Narrator: From the title of the act, the market economy seems to rage like Godzilla. I wonder what the market economy is going to attack. The entire world, perhaps. In Acts 1 and 2, however, Shinzuki, Majime and Morimori talked chiefly about problems in mathematical theories. This interpretation seems consistent with what the poet said. Accordingly, I suspect that the market economy here means market equilibrium theory. If so, market equilibrium theory might seek its revenge upon game theory. This will be interesting.

## Scene 1  Market equilibrium and social dilemma

[Majime appears in the office looking a bit gloomy]
Shinzuki  Majime, you don't look well, are you okay?
Majime  No, not really. Lately I have been having doubts about game theory, which worries me. Sir, may I ask you to listen to my problems?

Shinzuki  Well, it's quite normal for a serious researcher to have doubts about what he is doing. Actually, it is good for you.

Majime  You make fun of me, like always. I am talking seriously.

Shinzuki  Sorry. Let me listen to you seriously.

Majime  I have serious doubts about the relation between game theory and market equilibrium theory[1]. Among economists, or more narrowly, among the game theorists or mathematical economists, game theory is thought to be useful in order to study socio-economic problems, while market equilibrium theory, i.e., the theory of perfect competition, is hardly useful for such problems. I have been taking this attitude for my research and teaching. However, lately I think it might be slightly different as if a reversal, as you may call it, is also happening here.

Shinzuki  What? A reversal of game theory and market equilibrium theory? It sounds very interesting.

---

[1] Here, the term "market equilibrium theory" is used rather than "general equilibrium theory". The latter is more common in the economics literature. See Act 1 for the reason why we adopt the former here. There are many standard textbooks on this theory. Here, we cite only one reference: Arrow KJ, Hahn FH (1971) General competitive analysis. Holden-Day, San Francisco.

| | |
|---|---|
| Morimori | Really? I want to hear it too. Mr. Majime, I can't wait to hear your idea. If game theory and market equilibrium theory are reversed, I should become a market equilibrium theorist, shouldn't I? |
| Majime | Now, both of you are flattering me while I'm honestly in doubt. Actually, however, I'm starting to think it might be interesting. |
| Morimori | So, would you please start, Mr. Majime? |
| Majime | Well, where should I start? Let me imitate your way, Sir. Morimori, could you summarize market equilibrium theory, please? |
| Morimori | Now, do you use me as an engine to get the discussion started? Usually you play that role, but for a change I'm glad to do it. I remember that in an undergraduate class, Professor Shinzuki told us: |

*"Theoretically speaking, the structure of market equilibrium theory consists only of three skeletons. Textbooks are full of small facts, but only those skeletons are related to God's thoughts. The rest are details."*

| | |
|---|---|
| | I was quite surprised by that, because all other professors were teaching many small facts in market equilibrium theory rather than the skeletons. |
| Majime | That is typical for Professor Shinzuki. You borrowed these words from Albert Einstein, didn't you, Sir? Einstein said, *"I want to know God's thoughts ... the rest are details."* |
| Morimori | Those radical words are also borrowed! Unbelievable. But often you pay a lot of attention to details in other people's seminars, Professor. |
| [Shinzuki is a bit offended] | |
| Shinzuki | I'm not saying, "Ignore the details" but I want to say, "Proper treatment of details is just a first requirement." A musician who reads notes incorrectly, a painter who sketches inaccurately, a scholar who thinks illogically, do you think such a professional has any existential value? |

82　Scene 1　Market equilibrium and social dilemma

Majime　　I think it's better to continue, Morimori. Could you first repeat these three skeletons of market equilibrium theory?

Morimori　I'll write them on the blackboard.

(1) Theory of consumer behavior (utility maximization given a budget constraint and market prices)
(2) Theory of producer behavior (profit maximization given market prices)
(3) Theory of market price adjustment

This is all about the skeletons that Professor Shinzuki mentioned.

Majime　　Please don't stop here, and explain them.

Morimori　Okay. I recall Professor Shinzuki explained them in the following way.

For (1) and (2) we suppose that the behavior of each consumer or producer doesn't influence the market prices. From (1), the demand function of each individual consumer is derived, and the market demand function is obtained by just summing up all the individual demand functions. From (2), the supply function of each individual producer is derived, and the market supply function is obtained by summing up all the individual supply functions.

Then (3) is as follows: The market price adjusts to eliminate shortages or surpluses. The equilibrium price is the one where quantity demanded just equals quantity supplied, and thus this may be expected as a natural one. We attribute this idea of price adjustment to Adam Smith, who called it "the Invisible Hand of God".

Shinzuki　Good! It would be better to consider "the Invisible Hand" as applied to all three, meaning that the whole economy is led to equilibrium, i.e., "predetermined harmony". It is Adam Smith's emphasis that even though every agent pursues for his own utility or profit, a harmony for the whole economy should lead to a good and balanced state of the economy.

|  | From the methodological point of view, you have to pay attention to the basic difference between the treatments of the market prices on the individual level and on the whole market level. |
|---|---|
| Morimori | Ah... I remember what you emphasized. The price is treated in a totally different way for the individual consumer or producer and for the whole market. This is the essence of perfect competition. Your emphasis is: It is the basic postulate of (1) and (2) that the individual consumer or producer takes the market prices as given and fixed for the moment of his utility or profit maximization. This postulate is derived from the large number assumption that there are many consumers and producers in the economy. The large number assumption justifies the fixed treatment of the market prices on the individual level. That is about it, I guess. |
| Shinzuki | You must have learned something else from other professors. |
| Morimori | Yes, I'll try to recall a bit more. Another professor taught the first and second fundamental theorems of welfare economics. The first theorem states that the resource allocation determined by market equilibrium is Pareto optimal, which means that there are no other allocations that make all the consumers better off[2]. |
| Majime | Pareto optimality should be described more accurately: no matter how the production technologies are used, or no matter how goods are exchanged, all the consumers cannot be simultaneously better off, in other words, when some are better off, some others are necessarily worse off. |
| Morimori | Yes, that is slightly more accurate. |
| Majime | I think it is a lot more accurate. |
| Shinzuki | Okay, okay, I should note the fact that Pareto optimality mentions nothing about equity. For example, even though one person monopolizes all the goods, Pareto optimality remains valid |

---

[2] Cf. Mas-Collel A, Whinston MD, Green JR (1995) Microeconomic theory, Chap.10. Oxford University Press, New York.

           as long as there is no waste in the production and resource allocation of the economy as a whole.
Majime     Sir, that is an important point but I don't think it is directly related to my problems. So, Morimori, please continue with the second theorem of welfare economics.
Morimori   Okay. The second theorem states that an arbitrary Pareto optimal allocation is achieved by a market equilibrium by a suitable redistribution of income.
           In class, I understood those theorems mathematically, but I didn't understand why they were given such big names as "the first and second fundamental theorems of welfare economics".
[Shinzuki turns to Majime who is looking troubled]
Shinzuki   Morimori summarized market equilibrium theory well. So, Majime, what are you troubled with?
Majime     It is related to Morimori's last comment. I wonder whether the first and second fundamental theorems have enough content to merit that appellation. I can prove these theorems easily, and yet I think that their socio-economic contents are slim. What they can handle is only a market economy yielding a Pareto optimal allocation. What they cannot handle are present-day socio-economic problems, especially, environmental problems.
Morimori   What do you mean by environmental problems?
Majime     Examples are global warming caused by excessive burning of fossil fuels and deforestation, air pollution caused by automobiles, and the expansion of the ozone hole caused by escaping Freon from air conditioners, etc.
Morimori   Does market equilibrium achieve Pareto optimality in the presence of such environmental problems?
Majime     No, it doesn't. It may deviate seriously from Pareto optimal allocations, but the first fundamental theorem states that market equilibrium is Pareto optimal. Thus, market equilibrium theory is incompatible with environmental problems. It can be concluded that market equilibrium theory is totally powerless to study environmental problems.

Morimori   But game theoretical research has been done on such environmental problems. The Prisoner's Dilemma or the Social Dilemma is regarded as useful for the study of environmental problems. In general, the resulting outcomes in games may not be Pareto optimal. Therefore, environmental problems can be studied by game theory[3].

Majime   You are right to a certain extent. In game theory, Nash equilibria are not Pareto optimal in most cases. This is why game theory is expected to be useful to study socio-economic and environmental problems.

However, Nash equilibria may deviate only slightly from Pareto optimality. In oligopoly theory, Pareto optimality might be approximately achievable for a large number of firms. On the other hand, present worldwide environmental problems are more serious: when the population increases, the opposite polar case to Pareto optimality might occur.

Though game theory is expected to do something for environmental problems, it remains only by the reason of elimination; namely, market equilibrium theory doesn't work and only game theory is left.

Morimori   It is better to be than not to be, even because only the others are eliminated.

Shinzuki   No, if it is bad, it shouldn't be, even though it is less bad than the others.

Majime   Again the two of you are starting puns. I should return to my problem. The fundamental theorems of welfare economics are good but not more than simple theorems. When they are called fundamental theorems, they are really mistreated. I think that the present socio-economic problems we are facing are very serious. Market equilibrium theory must be capable of treating these socio-economic problems.

Morimori   Mm... Mr. Majime, about the same thing, one moment you are being negative about what you are saying and the next moment

---

[3] See Act 2, Scene 2 for the Prisoner's Dilemma.

|  |  |
|---|---|
|  | you are being positive. I have no idea about the direction you are aiming at. |
| Shinzuki | That is why Majime is in serious doubt. |

[Shinzuki changes his voices to an authoritative tone]

When one feels that an extant theory is unsatisfactory for some aim, he shall meet the necessity of a new theory and try to construct it. Any theory has both positive and negative sides. It is built by focusing on certain aspects, by ignoring other aspects and by eliminating irrelevant factors. Unless a lot of aspects and factors are adequately ignored and eliminated, the resulting theory could be an ugly monster or an uninteresting toy, and would be useless in either case.

|  |  |
|---|---|
| Morimori | I want to see an ugly monster, rather than an uninteresting toy. |
| Majime | Haha, but what is your point, Sir? |
| Shinzuki | I don't want to talk about monsters or toys, but instead, I want to note that any good theory is constructed through adequate abstraction. On the one hand, it is inevitable for any theory to have many aspects unfaithful to reality. That is, factors eliminated in the abstraction process cannot be considered by that theory. On the other hand, thanks to such abstraction, we can analyze the phenomena we want to consider in the theory. |
| Majime | How is your comment related to my problems? |
| Shinzuki | I should paraphrase what I stated. Each theory as a whole has a lot of negative aspects and may be useless for an aim different from the original one. |
|  | In addition, when one tries to construct a new theory, no matter how new it is, quite a few parts are borrowed from extant theories. Therefore, you should examine positive and negative aspects of extant theories. In such a case, it requires a huge amount of labor, since you have to examine and reconstruct each part of the new theory borrowing from some extant theories in a suitable manner. Majime, you are troubled because you are wondering if you should start such a task based on market equilibrium theory and game theory, aren't you? |

| | |
|---|---|
| Majime | I think so. But the problem is still unclear to me. That is why I asked you to listen to my thinking. Please let me continue just a bit more the discussion about the relation between market equilibrium theory and game theory. |
| Shinzuki | Mm ... what about me acting pro market equilibrium theory today? I have a tendency of being negative about it, so I might not be able to put my whole heart into being pro, but I'll try my best. After your talk, I shall discuss certain epistemological and institutional aspects of market equilibrium theory as my own problems. |
| Morimori | So now the problem is whether I should be pro or contra. |
| Shinzuki | It is almost noon. Let's have something to eat before we start the discussion.<br><br>However, there is one more thing I should point out. In market equilibrium theory, there is a distinction between dynamic and static. A dynamic theory includes time, while a static one doesn't. People say, "a dynamic theory is more general than a static theory", but that is not always correct. |
| Morimori | Professor, you mean "the reversal of particularity and generality", which we discussed already at length the other day[4]. |
| Shinzuki | Did we already discuss that? Mm... by the way, what did I want to point out? Oh, yes, I remember. I was going to warn you that a static theory without a time structure doesn't express a situation in which a market exchange happens really only once. The economy, of course, is repeated, and a static theory describes a stationary state in the repeated situation. Time doesn't explicitly appear in the static theory, because the theory is focused on a stationary state. |
| Majime | Sir, your warning is already mentioned in several places. For example, Hicks' old work described explicitly the economy in such a way[5]. |

---

[4] Act 1.
[5] Hicks JR (1946, 2nd ed) Value and capital: an inquiry into some fundamental principles of economic theory. Clarendon Press, Oxford.

Shinzuki   Lately, however, many people in our profession think different. They consider a static theory as if the economy happens only once. I suppose they are influenced by the game theoretical set up where decision making is considered in a one-shot game. Such people don't deserve to be called "economists".

Morimori   You were supposed to act as pro, but you are already acting contra, Professor!

Majime   Mm... it is important to emphasize that time is included in a static theory. I suppose this will be related to the epistemological aspects you will talk about, Sir.

Shinzuki   I see. I should have waited a bit longer. Anyway, let's go for lunch. After that we will listen to what you would like to say, Majime.

## Scene 2   Market failure and widespread externalities

[The three return from lunch and start the discussion]

Shinzuki   Since I'm supposed to be pro market equilibrium theory, I shall argue the importance of the first fundamental theorem. However, it's impossible for me to rationalize the second theorem, no matter what sophism I use.

Let me repeat the first theorem: It states that if all the economic activities are done properly in the market, an efficient allocation results in the sense of Pareto optimality. For this theorem, we need the assumption that a large number of agents participate in the market as well as the assumption that each has many competitors. The large number assumption guarantees that each agent has only a negligible power to influence the market prices and takes the prices as given.

In fact, we need various institutional arrangements for the proper functioning of the market. First, the large number assumption is needed. Yet, it is more basic that private ownership should be guaranteed: Private ownership is legally protected from violation by others. Also, one is legally free to consume his own goods or to use them to produce in any way he wants.

| | |
|---|---|
| Majime | Yes, private ownership is very basic for the market, but it is a standard assumption. |
| Shinzuki | Now, I shall give a less standard argument.

The assumption of private ownership doesn't necessarily imply that a consumer or a producer may behave simply as a maximizer of utility or profit. For example, some social traditions may prevent each agent from being a maximizer of utility or profit. Society might have developed a tradition such as ostracism against some economic behavior. Thus, we need to design social institutions so as to eliminate such social constraints. |
| Majime | The large number assumption implies that any individual agent cannot influence the market price, right? In addition, you claim that each agent should be free legally as well as in reality, don't you? |
| Shinzuki | Exactly. Society consists of a large number of people and there should be no traditions to constrain individual economic behavior. Such a society must be analogous to a large city. One characteristic of a large city is that each individual being is negligible and ignored, which is sometimes called alienation. This is a negative description of that characteristic, but simultaneously it may be regarded as a positive aspect of a large city. That is because each individual is free in the sense that society doesn't constrain him at all. For this reason, a large city absorbs young people from the countryside, but many of them will be alienated in the end. |
| Morimori | I have no clue if you are pro or contra. |
| Shinzuki | By its nature, the market economy has positive and negative sides.

The first fundamental theorem has an implication for the design of socio-economic institutions: When the market functions properly for a given objective, economic activities relevant for the objective will be optimal in the sense of Pareto. For this reason, economists claim that any socio-economic problems should be left to the market if possible, rather than implement- |

ing other designs. For this reason, privatization or decentralization of a national company may be encouraged.

Thus, the title of the theorem represents its contents well.

Morimori  I see, the title is not outrageous.

Majime  Morimori, you shouldn't be deceived so easily by Professor Shinzuki! His discussion contains the essence of the problem, but he is cleverly cheating and giving the affirmative conclusion for the first fundamental theorem. Indeed, most of his explanations are the same as those given by someone who acts as contra. This makes it easy for me to act as contra.

Morimori  Uh ... Mr. Majime, what are you unhappy about?

Majime  Professor Shinzuki's explanations are just fine. However, for the first theorem to be valid, we need the individualistic assumption that the utility or profit for each agent depends only upon his consumption or production as well as the market prices. This cannot be applied to an economy with the presence of environmental problems.

Morimori  That is why game theory is useful, as I already said.

Majime  Economists are inclined to have that conclusion, but I think market equilibrium theory has much larger possibilities in investigations of environmental problems.

Morimori  Mr. Majime, can you please be more specific?

Majime  Well, in the present society, we can't avoid externalities. In the $18^{th}$ century when Adam Smith was alive, the scale of the earth was almost infinite compared with human activities, but now the earth is overflowing with humans.

In a large city, we have many serious environmental and societal problems. These may be seen from the viewpoint of externalities. Thus, the first fundamental theorem should be stated more explicitly: "If there are no externalities, the resulting allocation is Pareto optimal."

Morimori  Mr. Majime, do you mean the assumption of no externalities should be stated?

| | |
|---|---|
| Majime | Yes, it should be. Then we should replace the title, "the first fundamental theorem of welfare economics", by a more moderate one. |
| Shinzuki | Let's forget for a moment about downgrading the title, and let's talk about externalities. |
| Majime | Okay, it is a good suggestion, Sir. In standard economics textbooks, external economies and external diseconomies are discussed together with "market failure". Famous examples are "the beekeeper and the apple orchard" and "the cleaner and the bakery". But no serious environmental problems were discussed[6]. |
| Morimori | In "the cleaner and the bakery", the bakery shop releases smoke and the clothes hanging to dry at the cleaner's are dirtied by soot. This is called external diseconomies. In one class, a professor said that the market doesn't function in such a case and this is an example of market failure. But I don't understand in what sense this is called "market failure". |
| Shinzuki | It is natural to have difficulties in understanding "market failure", since many problems are mixed up. We should specify "failure" as well as "the market". First, we should restrict the use of the word "market" to mean a competitive market, i.e., perfect competition, which functioning as a place to trade is summarized mathematically as competitive equilibrium. This specification simplifies our discussion a lot. Then, "the cleaner and the bakery" is a problem of bilateral negotiations but not a problem of the market. |
| Morimori | Does "market failure" mean that the competitive market malfunctions? |
| Shinzuki | Yes, it does. It would be better to divide this malfunctioning into two cases, which I write on the blackboard: |

---

[6] Cf. Mas-Collel A, Whinston MD, Green JR (1995) Microeconomic theory, Chap.11. Oxford University Press, New York.

(4) The market works as a place to trade, and its functioning is summarized as competitive equilibrium. Nonetheless, the resulting allocation does not enjoy Pareto optimality.

(5) The market does not work as a place to trade in the way expressed by competitive equilibrium.

Majime's claim to separate Pareto optimality from competitive equilibrium is relevant for (4): the notion of competitive equilibrium makes sense as expressing the functioning of a competitive market but may not guarantee Pareto optimality.

Morimori How about (5)?

Shinzuki "The cleaner and the bakery" is an example of (5). Negotiations may occur, but this isn't a competitive market. Only two people are involved and the right for clean air isn't priced in the sense of a competitive market.

Majime In economics textbooks, however, the argument continues: Once ownership is explicitly specified, we should leave the problem to direct negotiations by the people involved. Then, their negotiations result in a Pareto optimal outcome, which is called Coase's theorem[7]. This theorem emphasizes that even a local dispute may be solved in a market-like treatment. However, you claim that this isn't a problem of the market, do you?

Shinzuki That's right. We should separate problems of negotiations from those of the market economy. Problems of negotiations themselves have also some importance.

Majime I don't think so. Many present-day environmental problems are not local but more widespread. Since such widespread problems involve many people, direct negotiations between those people will be of no use to solve the problems. Coase's theorem can't be applied to such cases.

Shinzuki I agree with you: I think that Coase's theorem itself is cheap and poor. However, there are many serious problems even

---

[7] Cf. Glahe FR, Lee DR (1981) Microeconomics, Chap.13. Harcourt Brace Javanovich Inc, London.

among local environmental ones. In a problem like the Minamata Mercury Pollution[8], a large single company as a polluter stands against numerous small households as victims. This is still a local problem. In such a case, the negotiations between the people involved can't be fair at all. A large company has an overwhelming advantage over small households: A large number of small victims can hardly organize themselves, while a large company easily separates the victims and disables their cooperation.

By the way, Morimori, have you ever heard the name Tolstoi?

Morimori   Tolstoi? I think so. He is a Russian revolutionary, isn't he?

[Majime, laughing at Morimori]

Majime   Hahaha, no! He was a famous Russian novelist in the 19$^{th}$ century.

Morimori   Bu... you are knowledgeable and I'm silly, as you know. But the textbooks on economics or the papers on game theory I have ever read mention nothing about literature. Would literature be useful to study economics?

Majime   Sorry, you are right. Sir, why did you bring the conversation to Tolstoi?

Shinzuki   I should be sorry. I had no intention on checking on Morimori's knowledge of literature nor do I want to argue that literature is useful for economics or social sciences. I only wanted to quote some of the words of Tolstoi. The opening sentence of Tolstoi's *Anna Karenina* is:

"*Happy families are more or less like one another; every unhappy family is unhappy in its own particular way*"[9].

I apply this sentence to the problem of negotiations between the large single company and the numerous small households.

---

[8] A chemical plant caused the Minamata disease by polluting the Minamata bay area in Kyushu, Southern Japan, from 1953 to 1960.

[9] Tolstoi L (1912, original 1873-'76) Anna Karenina. Translated by Townsend RS, Vol.1 JM Dent & Sons LTD, New York.

Though mercury is a single cause for unhappiness of victims, it has created many forms of unhappiness since they may differ for each household. The unity of those diverse unhappy families cannot be compared to a unity of a few people who pursue the immortal existence of the company.

[Shinzuki takes a Kabuki[10] style posture]

*Notwithstanding the fact that to bring together contented people is very demanding, to unite many discontented people is out of the question.*

When the negotiations are left to the people involved, without a special leader or special supporters for the victims, the large company would cover up the pollution problem by gaining the consent from the victims with a minimal compensation.

Morimori    Is that the application of Tolstoi's theorem? Ah... literature is useful for economics.

Majime    A nice story. But now I would like to discuss externalities in relation to market equilibrium theory.

Morimori    Okay, I understand local externalities, but not widespread externalities. Mr. Majime, could you please explain the latter?

Majime    Yes. By "widespread externalities", I mean that externalities influence a large area and many people, but that any individual agent has negligible influence on the level of externalities.

The air pollution in a large city is an example of a widespread externality. The exhaust gas from the car of each person can be negligible compared with the total air pollution. When it is serious, each person experiences health problems. But nobody regards the exhaust gas from his own car as the direct cause for pollution.

Morimori    I see. Widespread externalities are also problems in a large city.

---

[10] Kabuki is "a (traditional) popular drama of Japan characterized by elaborate costuming, stylized acting, and the performance of all roles by male actors", Random House Webster's College Dictionary. (1990), Random House, New York.

Majime   They are quite relevant in a large city. In fact, they are not particular to a large city but also are related to many environmental problems. Here, I would like to emphasize the fact that widespread externalities have a structure parallel to the treatment of the market prices as given in a competitive market[11].

Morimori   The relation between each person and widespread externalities is parallel to the relation between each agent and the market prices in the competitive market, isn't it?

Majime   Exactly. Widespread externalities are directly and indirectly related to the market economy. In contrast to local externalities, widespread ones can't be the objects of negotiations, since too many people are influenced passively but each has active influences only at a negligible degree.

Notice that when widespread externalities are present, a competitive market may keep its functioning as a place of trading. In this case, "market failure" may occur only in the sense of (4) on the blackboard. It may happen that air pollution is severe in a large city but the people may not change their economic behavior. That is, the actual utilities for people drop, but their economic behaviors are not influenced very much by air pollution.

Morimori   Is it possible that utilities change but behaviors don't?

Majime   Yes, it is. Economic behavior is determined by the ordinal preferences over the economic variables that each person can choose, but may be independent of the pollution level. Nevertheless, the pollution may decrease the utility level itself.

Shinzuki   In such a case, a resulting allocation may be invariant with or without widespread externalities. The same allocation enjoys

---

[11] For the relation between widespread externalities and market equilibrium theory, see Hammond PJ, Kaneko M, Wooders MH (1989) Continuum economies with finite coalitions: core, equilibria and widespread externalities. Journal of Economic Theory 49: 113-134, and Kaneko M, Wooders HM (1994) Widespread externalities and perfectly competitive markets: examples. In: Gilles R, Ruyes P (eds) Imperfection and Behavior in Economic Organizations, pp 71-87. Kluwer, Amsterdam.

|  |  |
|---|---|
|  | Pareto optimality in the absence of externalities, but becomes terribly bad in their presence. This implies that the resulting outcome in the market economy may be far away from Pareto optimality in the latter case. |
| Majime | Exactly. This is the "market failure" in the sense of (4). |
| Morimori | But I previously asked if it is a game theoretical problem. I don't see why we should return to market equilibrium theory now. |
| Majime | The reason is that the economic activities of people in the market cause such a social dilemma. Also, the market often amplifies a social dilemma or widespread externalities. |
| Morimori | Uh... what do you mean by saying that the market amplifies a social dilemma or widespread externalities? |
| Majime | As an example of the amplification of externalities through the market, we can consider "the Tragedy of the Commons"[12]. Consider the situation where a lot of fishermen go fishing on a certain fishing ground. Each fisherman catches an amount negligible to the total amount of the fish resources. Thus, the more each catches with his own boat, the more profits he would get. But if all fishermen behave in this manner, the fish resources will be soon exhausted.<br><br>This is the basic structure of "the Tragedy of the Commons". The externalities here are widespread in the sense that the fishing grounds are large enough for each fisherman. Many other resource problems have a similar kind of structures. |
| Morimori | But you don't mention the amplification through the market. |
| Majime | Up to now, I gave the standard explanation of "the Tragedy of the Commons". Now I will address the problem of the market. Suppose that the fish resources have decreased. How does this affect the fish price on the market, Morimori? |
| Morimori | The price goes up since the supply decreases. Mm ... the price increase depends upon the shape of the market demand function. |

---

[12] Hardin G (1968) The tragedy of the commons. Science 162: 1243-1248.

| | |
|---|---|
| Majime | The price increase depends particularly upon the price elasticity of the demand. Since it is a food product, the price elasticity must be small. |
| Morimori | It means that when the fish price increases, the demand for fish doesn't decrease a lot. |
| Majime | The contrapositive of your statement is more standard. |
| Morimori | Ah... when the supply of fish decreases, the fish price will increase a lot to balance the demand and supply, because food has a small price elasticity. Is this correct? |
| Majime | Yes, it is. In the beginning, the price was cheap because a lot of fish could be caught. Over-fishing gradually decreases the total amount of fishery, while the fish price increases a lot. Even when the fishery amount decreases, the profits could be more than before since the price increases a lot. Thus, the fishermen continue to fish, and finally the resources will be exhausted. |
| Morimori | Okay, I understand the amplification of externalities through the market.<br><br>Then, the animals of which the import and export are forbidden under the Washington Treaty are a good example, aren't they? |
| Majime | Yes, they are. Even though hunting or trading is forbidden, the price on the black market is quite high. Hence, some people poach those animals and they become rarer. Then the price will rise a lot, which makes the hunters continue to poach those animals. In the end, they will be extinct. We have many examples like this. Behind those environmental problems, there are human economic activities and the market amplifies them.<br><br>In market equilibrium theory, even now researchers are continuing mathematical refinements and generalizations of the existence theorem of market equilibrium and of the first or second fundamental theorem of welfare economics. |
| Morimori | Mr. Majime, do you claim that the present environmental problems should be addressed by market equilibrium theory? |
| Majime | Yes, I do. To capture the environmental problems we have discussed, we need to understand their economic backgrounds cor- |

|||
|---|---|
| | rectly. Market equilibrium theory is now required to capture them. |
| Morimori | I'm starting to understand what you want to say, Mr. Majime. |
| Majime | Market equilibrium theory should be important as a theory to analyze the present environmental problems. So, we should downgrade the first fundamental theorem of welfare economics as one theorem in the case of no widespread externalities. |
| Shinzuki | I agree with you entirely. The market economy is assaulting the whole earth through environmental problems. Market equilibrium theory is supposed to analyze such problems, but it is progressing only in mathematical refinements and generalizations. Environmental problems laugh at market equilibrium theory. |
| Morimori | Now I understand your doubts well, Mr. Majime. But what are you really troubled about? Your explanation is consistent from beginning to end. Why don't you study environmental problems through market equilibrium theory, as you would like to? |
| Shinzuki | That is his problem. To do such research, he should go against the mainstream of the present economics profession. He isn't sure at all if he will obtain results from that new project. It will be a big challenge for him even when he believes he is going in the right direction.<br><br>Majime has been successful in our profession with his research on game theory. I think he is troubled about whether to take the risky challenge of building a new market equilibrium theory, doubting that he might get immediate results. |
| Majime | Exactly. What should I do in such a case? |
| Shinzuki | I think you said once, |

> *"Whatever might be expected, we should quest for truth. This is the duty for those gifted with noble minds, and moreover this is our choice even where there is no other choice."*

|||
|---|---|
| Morimori | Indeed, he said so. But, Professor, I heard that when you were young, people expected a lot from you. Are you, like Mr. Ma- |

|            | jime, challenging a big problem and still waiting for results to come? |
|---|---|
| Majime | Morimori, don't be so impolite towards your professor. |

[Shinzuki, shrugging his shoulders]

| Shinzuki | Let him talk like that. One day I will get my revenge on the two of you, you just wait! |
|---|---|
| | This was a long discussion. Let's have some tea now. After that I have to take care of some administrative issues, so let's continue the discussion after 4 o'clock. |

## Scene 3  Epistemological consideration of perfect competition

[Shinzuki slowly starts to talk]

| Shinzuki | Well, since Majime finished his talk, it is my turn now. |
|---|---|
| | The central notion constituting market equilibrium theory is "perfect competition". This notion is interesting particularly from the epistemological and institutional points of view, since it contains various seemingly contradictory aspects. I would like to discuss certain consequences from those aspects. Nevertheless, I should start with a warning about the nature of the following discussions. |
| | After the collapse of the Soviet Union in 1991, the political idea of "free competition" has become dominant in the world, though it is used almost interchangeably with "perfect competition". This idea is prevalent not only in the US, but has also been prevailing worldwide. For example, a lot of economists consulting at the World Bank or the IMF studied economics at graduate schools in the US. Their thoughts are based on free competition they learned at graduate schools. Through their work, the idea of free competition affects the rest of the world. |
| Morimori | Good, economics has great influence. |
| Shinzuki | But we should be careful about such influences, since they may originate from a simplistic misunderstanding of perfect competition. In general, it is difficult to think about what scope and what limit a notion has. It is much easier to conduct research on |

|  |  |
|---|---|
|  | separate pieces and to escape into technical details. Especially for a large notion such as perfect competition, it is difficult to understand correctly its scope and limit. For those who are being taught, it is nearly impossible to grasp. |
| Morimori | Yes, yes, if a bad teacher talks only about the limit of a subject, then students don't want to study it. Any teacher says that the subject is useful. |
| Shinzuki | In general, students don't appreciate the limit of a notion. They take concepts, notions and thoughts they are taught simply as for granted, and those have formed their language.<br><br>In fact, this is a general problem not only for students but also for us. The scope of our thought is determined by language, and the language is determined by education. Our thoughts and ideas are subject to a lot of falsities. Moreover, we have a tendency to modify observations in a manner comfortable for ourselves[13]. Thus, we should be very careful about acquired concepts, notions and thoughts. |
| Majime | I understand your warning, but it is applied to us in general. Does it have an implication in particular to the education of graduate schools in the US? I'm wondering because I did my graduate studies there. |
| Shinzuki | Yes, we have to be careful particularly with the influence of economics education in the US. Of course it is a matter of degree.<br><br>In the US, the beliefs of "individualism" and "free competition" are dominant. In education as well as research, both students and teachers are exposed to harsh competition. This fact has both good and bad aspects. It gives pressure to obtain results quickly, and makes people have a tendency to value myopic success in the competitive society.<br><br>In such a circumstance, it is quite difficult for people to think carefully about the notion of perfect competition, though it has great influence worldwide. |

---

[13] Act 2, Scene 4.

| | |
|---|---|
| Majime | In my graduate studies, indeed, we never discussed the limit or scope of perfect competition. But one couldn't write a paper anymore if he thinks about such limits. Mm... this "one couldn't write a paper" may be a sign of being affected by the competitive society. |
| Shinzuki | I think so. You are already quite polluted with the American beliefs. However, among the researchers I know, you can find more people in the US than in Japan, who are independent from competition for academic success. But we should return to our original problem, that is, epistemic and institutional aspects of perfect competition.<br><br>If economic theorists spend all their energy into refining and generalizing technical details but ignoring conceptual examinations of market equilibrium theory, then, as a result, there would be no meaningful development. It is even worse that the simplistic understanding of perfect competition or free competition affects actual societies and may even destroy them.<br><br>Therefore, market equilibrium theory should be conceptually examined more carefully. This helps us understand the hidden components or logical implications from perfect competition. Then, using such knowledge, we will be better able to prevent the establishment of social and economic institutions yielding negative effects on society. |
| Morimori | It all sounds very difficult, but I feel that you are aiming again at some reversal. |
| Majime | Sir, you are talking about very important problems for our choice of future research subjects. Please don't turn things upside down again. |
| Shinzuki | Okay, okay, I will be serious.<br><br>First, I should point out that my following argument is not about the target phenomena of market equilibrium theory. I will consider what may happen with society as a whole if we pursue to design a perfectly competitive society. My argument may lead us to an extreme society, but the real society might move |

in such a direction if we follow blindly the notion of perfect competition.

I will start with the epistemological aspects of perfect competition, but with what should I start concretely?

Morimori   Professor, before your long discussion, I want to ask about one problem I noticed. When we talked about the Konnyaku Mondo the other day, the assumptions of perfect information and complete information in game theory were mentioned[14]. Afterwards I was reading an article about perfect competition, and I noticed one epistemic problem of perfect competition.

Shinzuki   What is it?

Morimori   It was written that the assumption of perfect information is needed in addition to the assumption of a large number of agents. I don't understand this perfect information assumption at all.

Shinzuki   Ah... that is a good observation. It is related to what I will talk about. Let me start with it first. I guess that the article didn't specify the term "perfect information".

Morimori   No, it didn't specify that assumption at all. If it is the same as perfect information in game theory, then each agent could observe precisely the trades of all agents after each period.

Majime   That is the interpretation of perfect information in the sense of extensive games or repeated games, isn't it[15]?

Morimori   However, perfect competition includes the assumption of many agents. Isn't it strange that every agent can observe precisely the outcome for each of many agents?

Shinzuki   Yes, it is inadequate to make that assumption. Here, it is relevant to mention the two game theoretical results called the *Folk Theorem* and the *Anti-folk Theorem*.

As you know, the Folk Theorem states that in a repeated situation of a game with perfect observation, any outcome is possi-

---

[14] Act 2, Scene 2.
[15] Strictly following the terminology of the theory of extensive games, this is not the assumption of perfect information but of perfect observation.

bly sustained by some Nash equilibrium[16]. A situation described by the theorem is similar to some form of ostracism that has been developed to prevent any individual deviation from the social tradition. This result depends upon the assumption of precise observation of each single player's action, since ostracism couldn't work without this assumption. The assumption of precise observation may be adequate only for a game with a small number of players. In this sense, as Morimori doubted, the assumption of perfect information for competitive market totally differs from the game theoretical assumption of perfect observation for a repeated situation.

Morimori  I understand what you say about the Folk Theorem. Then, does the Anti-folk Theorem talk about the other case with many agents? Must it be true that no agents observe precisely the actions of the other agents?

Shinzuki  You are right. The Anti-folk Theorem states the opposite result under the assumption of no precise observations, i.e., each single agent can observe only the aggregate information. Under this assumption, the information about any single agent is ignored and his behavior cannot be constrained by any form of ostracism. Therefore, each can behave freely. Technically speaking, a Nash equilibrium outcome of each period is only attained in this repeated situation. This case corresponds to a perfectly competitive economy[17].

Morimori  But I recall, the other day, you said that the Folk Theorem is a horribly terrible theorem, didn't you[18]?

Shinzuki  Yes, I did. In its standard form, each agent plans from the *ex ante* viewpoint. From this, the Folk Theorem contains serious conceptual difficulties. However, you could understand it in a restricted manner: A social custom happens to take the form of

---

[16] For details about the Folk Theorem, see Osborne MJ, Rubinstein A (1994) A course in game theory, Chap.8. The MIT Press, Cambridge.

[17] For details about the Anti-folk Theorem, see Kaneko M (1982) Some remarks on the folk theorem in game theory. Mathematical Social Sciences 3: 281-290.

[18] Act 1, Scene 4.

penal controls such as social ostracism for deviations from the accepted social tradition, when people can observe the behavior of each individual agent. In this case, such a social tradition corresponds to a possible history of that society. It may take many forms. This is the interpretation of the Folk Theorem from a positive point of view.

The Anti-folk Theorem should be understood in a similar restricted manner. In a large city, only a few people may observe the behavior of each single individual. In such a case, penal controls don't work effectively. That is why every individual can be free in a large city. This is what the Anti-folk Theorem describes. Also, the informational assumption for the Anti-folk Theorem is exactly the information assumption for a perfectly competitive economy.

Morimori   I understand well neither the Folk Theorem nor the Anti-folk Theorem, yet. I will have to read some related material. Will you please explain it again in detail later?

Shinzuki   I will be glad to do so, but together with the conceptual difficulties.

Morimori   Okay, I see. I will read the material first.

The assumption of perfect information in market equilibrium theory isn't the same as perfect information in game theory, right? Then, let me compare it with the other assumption, i.e., complete information, in game theory. This means that each agent has complete knowledge about the utility functions of all the agents of the economy and of the production functions of the firms. But wait, for an economy with a large number of agents, this assumption is far away from reality, even further than the assumption of perfect information in game theory.

Shinzuki   But some researchers understand perfect competition in this manner, though it is definitely inadequate. In static theory, the economy is not played just once but is played repeatedly and a stationary state is considered, as I pointed out at the end of the discussion this morning.

Suppose that the economy is played only once and also that each agent plans his economic behavior as if he computes the entire market equilibrium. In such a case, we need to assume that each agent knows the structure of the whole economy to obtain his consumption or production plan as part of the entire equilibrium.

Morimori   Does each agent calculate the entire equilibrium to obtain his own consumption or production plan? Is he doing such inefficient thinking?

Shinzuki   Of course, this way of thinking has arisen from a stereotyped understanding of market equilibrium theory. Economists who think like this don't understand the process of abstraction to reach the mathematical formulation of market equilibrium theory. We cannot take this way of thinking seriously. Nevertheless, some textbook explanations of rational expectation theory are even an extension of this stereotyped thinking.

Morimori   So, what is the assumption of perfect information for perfect competition?

Shinzuki   In fact, a long time ago I was puzzled about that assumption for market equilibrium theory. Finally, I found it meant only that every economic agent knows the market price and the quality of the goods[19]. This is sufficient for each agent to maximize his utility or profit under the market price.

Morimori   Is that all? The knowledge of the market price and the quality of the goods is not really an important assumption. In particular, we may have some knowledge when the economy is repeated several times. Maybe there was no need to state that assumption again. Or am I mistaken?

Shinzuki   Well, you are right, more or less. If the economy is repeated, the assumption of perfect information in that sense holds approximately. We don't need to say that it is a necessary condition for perfect competition.

---

[19] Cf. Hayek FA (1964) The meaning of competition. In: Individualism and economic order, Chap.5. Routledge & Kegan Paul LTD, London.

| | |
|---|---|
| Morimori | Does that mean that perfect information isn't a useful assumption, Professor? |
| Shinzuki | Yes, you are right. |
| Morimori | I was right to think the assumption of perfect information is strange. |
| Shinzuki | In fact, I wanted to argue that the assumption needed for perfect competition is completely opposite from the perfect information assumption in game theory. One of the informational assumptions for perfect competition is that the individual behavior doesn't influence the whole economy through information. In other words, people observing the behavior of one single agent are negligible relative to the whole economy. This is called the assumption of *informational anonymity*. |
| Majime | Let me recapitulate in my own words. If the behavior of one certain individual can be observed by a lot of other people, he may eventually influence more people. He may even influence the market price. For example, a very famous person may change the society through information. This is contradictory to the assumption of perfect competition.<br><br>On the contrary, if few people see the behavior of one individual, it is impossible for this individual to influence the entire society. In this case, one is not constrained by society and can behave freely. Not only legal but also actual freedom is guaranteed for each individual. Is this the reason why you compared a perfectly competitive economy with a large city, Sir? |
| Shinzuki | Exactly. In perfect competition, every agent is anonymous qua information. Not many people notice him and the people whom he meets are only a negligible part of the economy. He has the freedom of not being observed by others, or negatively speaking, he is alienated. |
| Morimori | Uh... everything has both positive and negative sides. |
| Shinzuki | Morimori, can you give a real world example of a person who doesn't enjoy informational anonymity? |
| Morimori | It has to be somebody who is not anonymous. It must be a famous person, because he is observed by a lot of people. For ex- |

|  | ample, may he be the major league player Ichiro or the novelist Michael Moore? |
|---|---|
| Shinzuki | I was thinking of Hulk Hogan or Michael Jordan but nowadays Ichiro and Michael Moore seem to represent the celebrities. |

Shinzuki: In fact, the existence of celebrities is incompatible with perfect competition. As already stated, one requirement is that every agent is anonymous qua information relative to the whole economy. Also perfect competition doesn't allow the existence of rich people. Rich people may deny anonymity and also they might become richer and richer. Eventually they might be able to directly control the price. However, a lot of researchers have already addressed the problem of rich people. Therefore, I won't discuss it any further.

Morimori: Could you please state your conclusion?

Shinzuki: In fact, I would like to emphasize that we strengthened the large number assumption of agents for perfect competition by adding that each agent has many competitors. This additional part may be regarded as almost included in the large number assumption, but, logically speaking, it is an independent assumption.

[Morimori becomes irritated and impatient]

Morimori: To me it sounds the same as the assumption of many agents. When will you tell us what you really want to say?

Shinzuki: Okay, okay, my conclusion is simply that perfect competition is incompatible with our traditional culture and denies many of our values.

For example, consider arts. Art relies genuinely upon an individual and is truly personal. Excellent art is even more personal, and its rare existence has value. For example, think of Mozart or Bach as composers, or Cezanne or Van Gogh as painters. In academia, the same is true. Something really valuable is made only by a handful of people.

Morimori: In literature, we can think of Tolstoi or Michael Moore?

Shinzuki   That's it!
Please recall the informational assumption of anonymity for perfect competition. This assumption doesn't allow for the existence of outstanding people with salient originalities. Although individualism or original personality is regarded as valuable in free competition, it is nothing more than something substitutable by other people if they try to compete hard enough. In other words, that is what is meant by the assumption that each agent has a lot of competitors. Accordingly, genuinely original art or saliently particular achievements in academics shouldn't be allowed in perfect competition.

As already stated, the traditions of society are not compatible with perfect competition, because social traditions constrain the actual freedom of the individual. Thus, traditional virtues such as friendship, sincerity, trust, diligence etc. are not respected in perfect competition.

Majime   The freedom of economic behavior for the individual is legally guaranteed and in addition he is free from social traditions. In this sense it is an extremely free society. On the other hand, saliently original art or academics are not allowed in this soci-

|          | ety, right? However, if this guarantees social equality and efficiency, isn't that a small price? |
|----------|---|
| Shinzuki | Majime, then tell me what values can an individual have in a society with no good art nor academics and with no respect for traditional virtues. |
| Majime   | Mm... what values can an individual have in such a society? Individual value will originate from biological factors such as genes. Does that only leave us the quest for material satisfaction for the survival of ourselves and for the procreation of descendants? |
| Shinzuki | Indeed, only material satisfaction is dominant in such a society. But I find a positive aspect in that society. There, the utility of the consumer is purely physiological and includes no cultural factors. In this case, we can answer the troublesome problem of "what utility is" as "it is a physiological satisfaction". Thus, we solve that problem, hahaha. |
| Majime   | I give up. Economics is no longer a science of human society but is the same as evolutionary biology of the copulatory behavior of dragonflies. |
| Morimori | It sounds awful. I'm losing my motivation to study economics. |
| Shinzuki | You shouldn't be so discouraged. Here we discussed what would happen if we seek logical consequences of the basic assumptions for perfect competition.<br><br>I think that perfect competition is very useful both as an institution as well as an academic notion unless you regard it as the categorically dominant principle for social issues. Therefore, you should consider perfect competition not as a notion for the general theory but as one for a particular theory. |
| Morimori | Is that true? I don't believe you anymore, Professor. |
| Shinzuki | It's about time for me to go shopping with a small budget. Let's discuss the social and institutional aspects of perfect competition tomorrow morning. Both Majime and I have classes early in the morning, so how about starting at 10:30? |
| Majime   | That's fine with me. Well, I have to prepare for tomorrow's lecture. |

Morimori   I have nothing special for tomorrow. But I'm feeling blue.

## Scene 4   Institutional consideration of perfect competition

[Shinzuki and Majime have finished their lectures and return to the office]

Morimori   Good morning. How were your lectures?

Majime   Mine was the same as usual. How about yours, Sir?

Shinzuki   Good morning. My lecture? I couldn't reach the subject I wanted to cover today. If I don't teach my assigned task on calculus, other professors will complain. I shouldn't digress too much from the subject.

Morimori   What did you discuss in your class?

Shinzuki   I gave the $\varepsilon - \delta$ definition of continuity of a function. I asked the students if they understood. Some answered they didn't understand it at all and requested me to explain it intuitively. So, I started considering what intuition is, and then my consideration went in the direction to conclude that economists don't treat intuition as a scarce commodity, since economists always use their intuition as a reference point to make a judgment. Don't you think so?

Majime   Sir, do you want to digress here, too? We should continue yesterday's discussion.

Shinzuki   Right. But what is today's problem?

Morimori   Mm... Professor, don't you remember that strange conclusion we reached yesterday? I'll help you recall the problem.

Yesterday, we started discussing market equilibrium theory. Mr. Majime acted as contra and discussed economies with widespread externalities. Professor, you acted as pro, but actually almost as contra according to Mr. Majime. Anyway, we finished the discussion on the problem that Mr. Majime raised, and then, Professor, you started talking about one epistemological aspect of perfect competition. Your argument almost concluded that perfect competition led us to an animal society. Today we will consider institutional aspects.

## Act 3  The market economy in a rage

Shinzuki   I see, today we are supposed to consider institutional aspects. But my mind is still inclined to think about what intuition really means. I'm not ready for the discussion. Majime, can you give some keywords to remind me of the main point?

Majime   Yesterday, when we were interpreting the first fundamental theorem of welfare economics, the term "private ownership" came up. It may be a keyword for institutional aspects of perfect competition.

[Shinzuki starts slowly talking with a rap rhythm]

Shinzuki   I see, icy, that is private ownership certainly.
I don't know what it is all about uncertainly.
My head is malfunctioning with certainty.
Shouldn't I call it a thought failure or a sort of failure?
A bit changing, I have a make-up failure or a market failure
Aha, aha, we discussed it yesterday, didn't we?
Market malfunctioning or make-up malfunctioning isn't it?
Majime gave the keyword private ownership, didn't you?
Private ownership is functioning improperly.
Okay, ok, my head starts functioning finally.
Let me warm up a bit more, a lot more.
Tell me what private ownership is, tell me more.
Someone violates the ownership certainly.
What makes private ownership function properly?
I see, icy, I should ask the police to protect it, shouldn't I?
Yah, yah, it is the duty of cops, isn't it?
We never ever discussed cops yesterday, did we?

[Shinzuki starts to talk more or less in a normal way]

Now I recall well what we have discussed. For some reason my intuition didn't work in the beginning. I need to play with words for a while to warm up and to make my head function properly. Thus, our intuition is truly a scarce commodity. That is why I discussed intuition in my class.

Majime   Let's start the discussions about private ownership before he starts to talk about intuition again.

[Majime, a bit like a rapper]

|          | In order to have private ownership to be respected, we need a police power, don't we? |
|----------|---|

|          | For that reason, the police provides the public service of keeping the law, don't they? |
|----------|---|
| Morimori | Mm… the rhythm doesn't sound nice. |
| Majime   | That is right. I shouldn't mimic such vulgar behavior. |
| Shinzuki | Anyhow, we should return to the story. The police are a public service. Finally we arrived at the point I wanted to talk about. |
| Morimori | Did you already reach your point? It is much faster than usual. I should ask you to talk always as a rapper. Then what is your point, Professor? |
| Shinzuki | The police services are needed to protect private ownership so as to effectuate a perfectly competitive economy. The government provides police services as one kind of public services. Behind all these public services, however, there are public servants, who actually provide them. What happens with their salaries? |
| Morimori | Of course, public servants need salaries. |
| Majime   | Morimori, Professor Shinzuki asks about the financial sources for their salaries. |
| Shinzuki | That's right. The financial sources are collected in the form of taxes levied on people. Thus, the government provides police services to protect private ownership, and at the same time, the government violates private ownership by levying taxes on people. That is, levying taxes is the denial of private ownership but its aim is to protect private ownership. Without taxes the police power couldn't be maintained and we will end up with a lawless society. In either case, private ownership would be denied. |
| Majime   | Mm… it is the paradox of private ownership. However, our starting point is perfect competition. For it, private ownership has to be protected, but it has to be violated by levying taxes. Contradictory factors are contained already in the notion of perfectly competitive economy. |

| | |
|---|---|
| Morimori | But if we suppose that the police take the taxes and then protect just private ownership, would this create serious problems? |
| Shinzuki | I think it would create serious problems. First, the police consist of humans and they consider their positions as well as their merits. Accordingly, the police have to be considered as acting actors in society. In this case, we should consider the government including the police and the tax department in a perfectly competitive economy. |
| | The existence of police might lead to the denial of the anonymity assumption that we discussed yesterday. In order to collect taxes correctly, the tax department should capture the income of each person. Accordingly, the police or the tax office needs to know exactly the economic behavior of each person. Thus, anonymity isn't fulfilled, and perfect competition wouldn't hold. |
| Majime | In yesterday's discussion, each agent has only negligible influences on the entire economy either directly or indirectly via information. Certainly, if there exists a central office in the police or tax department to have all information about the economic behavior of all people, then the anonymity assumption would be denied. |
| | However, even with the police and tax department, if the number of observers of each economic agent is small, then the anonymity assumption is still fulfilled, isn't it? |
| Shinzuki | Uhm... but we need the assumption that everybody is observed by somebody else. If we pursue seriously the possibility of "every single agent is observed by only a limited number of people", we would end up in a horribly terrible society. |
| | Imagine the following: A few people observe the economic behavior of each single person, but each of those observing people is now observed by a few others. We need to suppose here that the people who are observed don't know who are observing them to avoid collusive tax evasion. Also, as already stated, the police can't control the central office that collects all information. |

|          | We complete the description by constructing an informational network of observers and observees so that its mesh covers the entire society with no final receivers of information. In this case, everybody pays taxes correctly: if someone evades, the observers behind him expose him to the police. |
|----------|---|
| Morimori | Everybody is observed secretly by somebody else at any time. That is horrible. I don't want to live in such a society. |
| Shinzuki | Let's recall yesterday's final conclusion now. A perfectly competitive economy is incompatible with the existence of saliently outstanding people in arts or academics. Social traditions and cultural traits are also prohibited since they may limit the individual's economic freedom. To a great extent, in that society people are equal as animals of one species are more or less equal. |
|          | The society described by the conclusions of yesterday and today is, I think, quite similar the society described in George Orwell's *1984*[20]. Big Brother is watching everybody, but no one knows who is behind Big Brother. *1984* continues with the description of language control by deleting all words having critical and cultural meanings in order to eradicate critical and cultural concepts. Such concepts are unnecessary for material maintenance of the society. After the eradication of such critical and cultural words, no one will have doubts about the present social system. This is the method to implement a perfectly competitive economy. |
| Majime   | It sounds like an extremely controlled society. People compete freely for material utilities. At the same time, other social aspects like tradition, culture, arts, academics etc. are almost prohibited. Can we call this a human society? |
| Shinzuki | Yes, we can. If the members of the society are human, it is a human society by definition, hahaha. |
| Majime   | Well, that is true but… |

---

[20] Orwell G (1949) 1984. New American Library, New York.

Shinzuki  By the way, do you have any idea of how the life of an individual being would be in such a society?

[Morimori turns his head slightly]

Morimori  Mm ... first, everybody is lonely. He has only a few friends and even these few may be his competitors.

Shinzuki  People are lonely as in a large city.

Majime  According to your explanation, Sir, culture, arts and academics are sacrificed. Therefore, it is a culturally poor society by definition. Economic activities decline and the society becomes materially poor as well.

Morimori  Lonely and poor as well, what a miserable society it is.

Shinzuki  And because people are controlled by the desire to procreate descendants, the population increases steadily. The environmental problems that you discussed yesterday will get worse. The existence of leaders of environmental movements is, similar to the existence of rich people or celebrities, not allowed in such a society. Concepts like "leaders" or "environmental problems" may be already deleted from the language. In such a situation the average life expectancy is short.

Well, Morimori, could you please summarize the life of a being in such a society?

Morimori  Well, every individual would be lonely, poor, and miserable, and finally the life expectancy is short. What a horrible society it is!

Shinzuki  Thank you for summarizing. From the thorough investigation of the notion of perfect competition, we obtain the logical consequence: In such a society, the life of man is lonely, poor, miserable and short.

[Majime looks surprised]

Majime  That sounds similar to something I have heard somewhere.

Shinzuki  It is similar to Hobbes' famous lines describing *the state of nature* in *Leviathan*:

> *In such condition, there is no place for industry; ... no culture of the earth; ... no arts; no letters; ... and the life of man, solitary, poor, nasty, brutish, and short*"[21].

Majime    Bu ... but I believe that the state of nature in *Leviathan* describes a complete lawlessness society. Is it the same as where perfect competition leads? It sounds like an unnaturally forced conclusion.

Shinzuki    But wasn't it the two of you who helped me develop my argument, like usual?

Majime    Something is strange. I think that this time we have to discuss the problem more carefully.

Morimori    I feel the same. What is the problem? Mm... I feel as if you are obscuring our view by a cloud of logical arguments.

Majime    Ah... I see. The contents of the discussion were not a problem but your way of talking is strange.

Even leading us to the state of nature of *Leviathan*, you manipulated carefully your words and led the conversation forcefully to the state of nature.

I sometimes think you want only to wrap us up in your discussion. Making us silent by obscuring our view by a cloud of logical arguments differs from convincing us soundly. In the former case, you led us to a strange conclusion and we only feel as if you are constructing a strange logical argument again. Do you really want to convince us from your heart?

Shinzuki    Of course, it is my true intention to explain the argument and to convince you. As an atheist, I can't swear to Zeus like Socrates. Honestly, I should confess I sometimes lead you to a strange conclusion out of pure mischief.

Majime    This is a good opportunity to complain about other things. You always build your argument in a way that is favorable for you. We agree with each step of your argument but we are led to a

---

[21] Hobbes T (original 1651) Leviathan, p.253. In: Woodbridge FJE (1930) Hobbes Selection. Charles Scribener's Sons, New York.

strange conclusion, while we are left in the dark about whether the entire argument is really correct or not.

Morimori   That is quite similar to giving excuses to rationalize anything for one's own merit, which you dislike, Professor.
[Morimori imitates Shinzuki's talking]
   Therefore you shouldn't hasten yourself to a conclusion but instead examine each step in detail and then discuss the whole thing carefully.

Shinzuki   Hahaha, I lost this one! But do you really think I'm hastening the discussion? I will take your complaints seriously now.

Majime   I would like to return to the conclusion you reached a little while ago. Sir, do you really think the world nowadays is moving in such a direction?

Shinzuki   I don't think we should be so pessimistic because there are many factors deviating from perfect competition in the present world.
   For example, a few large companies dominate the market or try to control the entire economy by cooperating cleverly with the government. Since the information technology is progressing,

anonymity won't hold. The government or big companies start watching people, and eventually satellites will be able to watch all people. Privacy will be a dead concept, and the word itself may be deleted from our language. In addition, a few wealthy people or a few successful people will become more and more wealthy.

Those factors will prevent perfect competition from its realization. Thus, we need not to worry about it.

Morimori  That is as bad as a perfectly competitive economy. In either case, we are going to a horribly terrible world. Nothing can save us.

Majime  Hm ... what should we do? I would also like to ask about the possibilities of market equilibrium theory we discussed yesterday.

Shinzuki  Go ahead.

Majime  After all, do you think that market equilibrium theory is useless?

Shinzuki  No, not at all. My conclusion is the opposite. I think there is no other theory comparable with market equilibrium theory. However, you shouldn't consider it as a theory about the entire society but as one that is fit for certain economic phenomena, more like a special or a partial theory.

Therefore, I agree totally with what you said yesterday, Majime, about extending market equilibrium theory to incorporate widespread externalities and study it in relation with the economic activities and market economy.

Today we discussed what would happen when we consider market equilibrium theory as a theory covering the entire society. Then we might end up with a horribly terrible society. From now on we should give up the simplistic idea that one general theory can tell all but we should construct a theory that is applicable to a particular economic phenomenon.

Majime  I thought we reached a negative conclusion but you agree with my proposal.

|  | As for considering a special theory rather than a general theory, I will be very happy if we can construct a theory that works well in an adequate domain of applications. |
|---|---|
| Morimori | I agree too. I understand now that even literature is helpful in economics. |
|  | But, I have one question after listening to your argument, Professor. How can we discuss the applicability of market equilibrium theory? Maybe, we need a general theory that tells us to which economic phenomena market equilibrium theory is applicable and to which it is not. |
| Shinzuki | Hahaha, that is the second reversal of "reversal of particularity and generality". What should we do? |
|  | We discussed quite a lot yesterday and today. I'm tired. Shall we go and have something good to eat? |

Narrator: After all, we will be lonely and miserable in the future, independent of achieving perfect competition or not. What shall we do about that? Maybe, the writer Thomas Carlyle was right in that he called economics *the dismal science*[22]. Incidentally, in the beginning I imagined that the market economy would be raging like Godzilla but it turned out to be the monstrous fish Leviathan of Thomas Hobbes. I'm proud my intuition is not so bad. But, do we need a general theory or should we forget about it? I'm at a total loss.

---

[22] The term "dismal science" was first used by Thomas Carlyle in the context of economic supports for the emancipation of slaves but unrelated to Malthus's population doctrine. See "The origin of the term "dismal science" to describe economics" by Dixon R http://www.economics.unimelb.edu.au/ Tldevelopment/econochat/Dixonecon00.html

## Interlude 1  Clouds hanging over economics and game theory

Narrator: The setting is in the cafeteria near the institute. Morimori is having coffee in the cafeteria, and sees Majime outside in a hurry and in a bad mood. Morimori calls Majime and asks if anything happened with him. I am wondering how this setting is related to the title of the interlude. Will it rain on the stage? Let's listen to what Majime has on his mind.

Morimori   Hello, hello, Mr. Majime, please wait. You look angry. Are you alright?

Majime   Ah... Morimori, I'm still furious with K of our institute. I attended a panel discussion at the meeting of the Japanese Economics Association the other day. There, K talked about his terrible opinions. I have his handout here. Have you looked at it already?

Morimori   No, I haven't. But I heard Professor K said something terrible about industrial waste.

Majime   Mm, that's correct. Indeed, he talked about industrial waste. I think what he said is really bad. You should look at his handout.

[Majime has left, and Morimori turns to the audience]

Morimori   I wonder what made Mr. Majime so mad to say "really bad". Let me see what is written.

[Morimori starts reading the handout]

### Clouds hanging over economics and game theory

Game theory and market equilibrium theory have been regarded as quite different in recent developments. Market equilibrium theory is based on the concept of perfect competition from its inception and has a limited scope of target problems. It has been regarded as inflexible with a narrow scope, and also its research has been criticized as only focused on mathematical generalizations and refinements. On the contrary, game theory is regarded as useful in studies of various socio-economic problems, and game theory is actually used in many fields such as economics, political science, computer science, management science, sociology and even biology. Presently, game theory is expected to have further theoretical developments together with its applications to many fields.

Nonetheless, looking carefully at game theory and market equilibrium theory, we find only minor differences in their mathematical structures. A salient difference is that game theory is capable of putting many different interpretations to the same structures, while market equilibrium theory does not allow such interpretations. In fact, the freedom of game theory often leads to konnyaku mondos.

Market equilibrium theory has been regarded as being in a crisis for a long time. I think, lately, game theory also is in a crisis though in a different form. How is game theory in crisis? A huge number of papers repeat the same idea with different interpretations. These papers have been piled mountain-high and quickly weathered into industrial waste. Such industrial waste blocks the eyes of researchers from seeing the frontier of the field. Researchers do not think about what is beyond such piles. They may have already lost the concept of "beyond" in their thoughts.

My own contention is as follows: Market equilibrium theory is a successful theory. But it has reached one stage of completion, and is stagnating for the time being. On the contrary, game theory has also been facing a serious crisis without reaching a stage of completion. When one wants to talk about the future of game theory and market equilibrium theory, we should return to the origins of these theories. The initiators made a lot of trials and errors before their inceptions, and might have much wider views and scopes than contemporary researchers in these fields. Therefore, to look back at the origins would help us explore new possibilities. In this report I will consider the origin of game theory and what kind of development is required from now on.

I wrote this manuscript to present it within a given limited time. Thus, I have sacrificed some subtleties in expressions and some parts might sound too strong.

1. Although game theory targets social and economic phenomena, the origin of game theory was not in social sciences but in mathematics. The

origin can even be pinpointed to the third crisis of mathematics, which was caused by the paradox of Cantor's set theory found by Bertrand Russell in the beginning of the 20$^{th}$ century. This coincided with the time of the crisis in physics. These crises were closely related both in thinking process as well as in their histories. Many serious scholars were involved in these crises and were forced to rethink the existing thoughts as well as human reasoning from their very bases.

2. In order to overcome the third crisis of mathematics[1], three schools were founded in the mathematics world:
   (a) The Hilbert School of Formalism (also called the Axiomatic School)
   (b) The Brouwer School of Intuitionism
   (c) The Russell-Whitehead School of Logicism
   These schools had long, harsh and bitter disputes on foundations of mathematics in the 1910's and 1920's.

3. In the second half of the 1920's, the young Neumann was one of the representatives of the Hilbert School of Formalism. Around that time, the school aimed to prove the contradiction-freeness of the mathematical systems of those days, based on Hilbert's proof theory. This was supposed to rescue the entire mathematics from the third crisis of mathematics.

4. Hilbert's proof theory was a quest for the foundation of mathematics and became a theory of mathematical activities (reasoning) rather than a theory of contents of mathematics. In other words, it was a mathematical theory of mathematical behavior of the idealized mathematician. Inspired by this, Neumann initiated the theory of games as a mathematical theory of human behavior in society.

5. From the beginning, Neumann borrowed tools for his theory of games from the foundational disputes on mathematics. In 1932 (published in 1937), Neumann reproved his minimax theorem as well as the existence theorem of a market equilibrium using the fixed-point theorem of

---

[1] The first crisis in mathematics was brought about by the discovery of an irrational number in the ancient Greece. The second crisis occurred in the end of 18$^{th}$ century with the lack of the rigorous definition of a limit concept, where a lot of paradoxical results were found too.

Brouwer, who was the chief exponent of the Intuitionist School and the enemy of Hilbert in the foundational disputes.
6. In 1944, Neumann published together with Morgenstern their great work: *Theory of Games and Economic Behavior*.
7. The third crisis of mathematics and the crisis of physics required scientists to reexamine the existing ideas, sciences and theories from their very bases. These brought about various new thoughts. Many great theories and technologies of the $20^{th}$ century were born from these crises. Only some have remained successful but many others have been forgotten and became extinct.
8. The theory of games was born simply as one of many theories after the third crisis of mathematics. The theory of computation, which became the foundation of the present computers, also originated from the crisis. Neumann himself contributed a lot to the initiation of computer science. Based on computation theory, he exhibited profound thoughts for biology, neurology and sociology in his last work "self-reproducing automaton"[2]. However, these thoughts are still almost unrelated to game theory.
9. Neumann allowed the free use of probabilities in game theory. This enabled him to prove the minimax theorem. However, the concept of probability has remained as ambiguous and shaky as ever before.
10. Von Mises' frequentist theory of probability was a quest for the analysis of the concept of probability. This theory was forgotten and expelled by Kolmogorov's measure-theoretic probability theory. The reason for this is that Kolmogorov's theory is far more operational than von Mises' while ignoring conceptual and foundational problems of probability. However, the late Kolmogorov reconsidered the frequentist theory of probability from the point of view of computation theory in the 1960's[3].
11. In game theory after Harsanyi and Aumann, game theorists have fled to subjective probability, which is the probability concept of an "I" story.

---

[2] Von Neumann J (1966) Theory of self-reproducing automata. Edited and completed by Burks AW. University of Illinois Press, Urbana.
[3] Cf. Weatherford R (1982) Philosophical foundations of probability theory, Chap.IV. Routledge & Kegan Paul, London.

They did not separate thoughts from tastes. Subjective probability is claimed to have an axiomatic foundation, but it is just a different representation of subjective probability and utility in terms of preference relations based on tastes.

12. The research on mathematical logic as a theory of mathematical reasoning reached a peak with Gödel's incompleteness theorem (1931). His theorem meant the collapse of Hilbert's program of proving the contradiction-freeness of the contemporary mathematics. Neumann heard Gödel's results from Gödel himself at the Konigsberg conference (1930)[4] and withdrew quickly from proof theory.

    However, the real development of mathematical logic started in the second half of the 1930's.

13. Game theory has interacted little with other thoughts originating from the third crisis of mathematics. The extensive game theory handling interactions of information and behavior was new, but the present game theory is not very different from classical economics. Even for the handling of information, game theory remained in the classical set theory without using results from mathematical logic. The problem of human reasoning has to be the foundations for human behavior but is almost forgotten. The mainstream of the present game theory differs totally from Neumann's hidden intention to compete with Hilbert's proof theory.

14. The present game theory still lies within the original scope of Neumann. Until 25 years ago, the population of game theorists was sparse, and the scope was sufficiently large for them. However, as the number of game theorists increased, it has become overcrowded. Now, many are competing for marginal contributions. The concern of each is only the success in his academics. The critical situation of the world does not enter his eyes. The excuse of *"publish or perish"* has been repeated to rationalize this myopic behavior. It is exactly the same as economists or politicians who are concerned only about problems in front of their eyes, even though the earth is facing serious problems such as global warming and environmental destructions.

---

[4] Cf. Die Naturwissenschaften 18 (November 1930): pp.957-1083.

What should be done to overcome this crisis? First, let us stop making variations by reinterpreting or combining existing components. Such reinterpretations and combinations will only increase the piles of industrial waste. Second, we have to recognize the fact that we are in a crisis. The third crisis of mathematics taught that we should rethink the existing problems, concepts, theories and thoughts from their very bases. This is precisely the way to overcome the impasse of game theory.

I hope that bright young people will be resolute to face directly these foundational problems without being caught in superficial problems.

Morimori   What? This is not about industrial waste but just about old stories related to game theory. Does Professor K want to call unimportant papers industrial waste? Why was Mr. Majime so upset? Mm… maybe, because he thinks his papers are classified into industrial waste. This must be the case, ha ha. But … if his papers are industrial waste, so are mine. If that is the case, I will get mad too, since I'm working so hard.

## Interlude 2  Game theory in a crisis

Narrator: In this interlude, two new participants will appear: Jan Hammer who is visiting Ts university for 6 months and Oliver Otsuki who came to participate in the international workshop "Epistemic Logic and Game Theory". These two together with Kurai Shinzuki will discuss the crisis of game theory. This seems to be a continuation of the previous interlude. I myself don't understand the critique by K and still believe that game theory has been making steady progress. This time, I can hear opinions of people from other places, and perhaps, they have a different viewpoint. Now, it would be better to listen to their discussions to find out whether the present game theory is really in a crisis.

> [Hammer, Otsuki and Shinzuki have ordered their lunch at a Mexican restaurant near the campus, and are nibbling on a salad and nachos taken from the buffet]

Hammer    By the way, Kurai, I heard that in some meeting, K of this institute said something terrible like "game theory nowadays is a landfill of industrial waste". Do you know what he meant by that?

Otsuki    What, a landfill of industrial waste? He's certainly courageous or must be insane.

Shinzuki    It was in a panel discussion about microeconomics and game theory held in the meeting of the Japanese Economic Association in October. But I believe he used the expression of "a pile of industrial waste" rather than "a landfill of industrial waste".

Hammer    In either case, it is horrible to say such a thing, but I want to know what K really meant.

Shinzuki    I attended the panel discussion and also I read his handout. I think I understood what K tried to say. In the panel discussion, however, some of the audience got angry because they thought he was calling their work industrial waste. The panel discussion ended full of antagonism.

Otsuki    Antagonism sometimes makes the discussion productive, doesn't it?

Hammer    I agree with you but I'm not particularly interested in the antagonism itself, since it is simply emotional. I'm curious about the cause of the antagonism. Namely, in what sense K used the metaphor "industrial waste" instead of directly criticizing the present game theory.

Shinzuki    That is right. The antagonism was quite emotional and uninteresting. Regarding "industrial waste", I don't think he associated a particularly deep meaning with it. Game theory has been popular lately, but K thinks that no substantial developments have occurred and too many papers are written based on almost the same ideas.

Hammer    I agree with K that lately there have been no substantial developments in game theory. But still the metaphor of "industrial waste" is too much, isn't it?

Shinzuki    His critique has another aspect. May I reconstruct the other part?

Hammer    Yes, please.

Shinzuki    First, I must emphasize that it is K's critique, not mine.

| | |
|---|---|
| Hammer | Yes, I understand. Please continue. |
| Shinzuki | Okay. The researchers who write such papers aim only to publish their work in so-called first-class journals in our profession. They conduct research only for their personal success rather than out of academic interest. Our profession has become an industry to produce papers, where the main concern is the sale of those papers. The contents are secondary and often have no real value. K called such products industrial waste. |
| Hammer | He regards the game theory community as an industry and the papers as products of that industry. I understand why he used the metaphor of "a landfill of industrial waste". |
| Shinzuki | It was "a pile of industrial waste". |
| Hammer | I understand. |
| Otsuki | Okay, okay, it is horrible in either case. The game theory community is now quite large, so perhaps, this tendency is natural. Jan, have you noticed people acting in the way K described? |
| Hammer | Yes, I know many people who write papers in that way. Even among those regarded as respected researchers, there are quite a lot who aim only at publishing papers in so called first-class journals. It is even worse that while some of them know that research is going in a wrong direction, they ignore it and still continue to publish papers. I agree with K that our field is flooded with industrial waste of papers as well as researchers. |
| Otsuki | Jan, you are radical, too. I thought I knew the recent scene of game theory. But is the present situation really so bad? When did that start to happen? Is it a very recent trend? |
| Shinzuki | In fact, I have observed that trend for quite a long time. Game theory started to become popular in the beginning of the 1970's. It became really popular in the beginning of the 1980's. Lately, a lot of papers have been written claiming that game theory can explain this phenomenon, that phenomenon, and others. Some are claiming even that game theory is able to explain every social phenomenon. In reality, almost all are variations on the same idea. |
| Hammer | Thus, game theory is in a crisis, isn't it? |

| | |
|---|---|
| Shinzuki | Apparently that is what K wanted to say. |
| Otsuki | Game theory was considered as a new field long time ago, but 60 years have passed already after von Neumann and Morgenstern. It has been standardized and also industrialized. Following the terminology of Thomas Kuhn's *"The Structure of Scientific Revolutions"*[1], game theory has entered the phase of "normal science", and now a lot of people have been living as professional scientists with it.<br>Do you know Kuhn's *"Scientific Revolutions"*? |
| Hammer | I have heard about it. |
| Shinzuki | I read it many years ago, and I have now only a vague memory of it. |
| Otsuki | Let me explain Kuhn's *"Scientific Revolutions"*. It is one of my favorite subjects. I need a piece of paper to write down the evolutionary process according to Kuhn. Kurai, could you please pass me that paper napkin? I can write on it with this ballpoint pen. Thanks.<br>According to Kuhn, the evolution of a science occurs in the following order.<br><br>(1) Prevalence of a Paradigm<br>(2) Normal Science<br>(3) Finding of an Anomaly<br>(4) Crisis<br>(5) Scientific Revolution<br>(6) Adoption of a new Paradigm<br><br>Here, the term "paradigm" is used to mean "a way of thinking provided by the dominating science of a certain era". "Normal science" means a science gradually progressing in that paradigm. The term "anomaly" points at a phenomenon that doesn't fit in the paradigm. When an "anomaly" appears showing some serious inadequacy of the paradigm, that science will face a "crisis". In order to overcome the "anomaly", a new way of |

---

[1] Kuhn TS (1964) The structure of scientific revolutions. Chicago University Press, Chicago.

|  |  |
|---|---|
| | thinking is required, and then a "scientific revolution" happens. When such a revolution results in the acceptance and initiation of a new way of thinking, it will become a new paradigm and the whole process will start anew. |
| Hammer | A serious anomalous phenomenon needs to be found for a substantial development of a science, doesn't it? Of course, some researchers need to notice the critical situation of their field, because otherwise nothing would happen. |
| Otsuki | Yes, I think so. The fact of game theory facing a crisis means that some serious anomaly has been found and some game theorists like you notice it. |
| Hammer | No, I don't think so. Hm, my belly is growling. Our food still is not here. I'm starving. Should I be more patient? |
| Shinzuki | Yes, you should. Have some more nachos. |
| Hammer | Nachos save me! Thank you very much. |
| | I agree with K that game theory is in a crisis, but I see no anomalies in game theory. |
| Otsuki | That is interesting. You claim that game theory is in a crisis without facing a serious anomaly. That is a serious anomaly in Kuhn's paradigm *"Scientific Revolutions"*, isn't it? |
| Hammer | I'm not sure. At least, I doubt that game theory is facing a serious anomaly. |
| Shinzuki | I doubt it too. First of all, we should notice that Kuhn targeted natural science. There seems to be a big difference between natural science and social science. |
| Otsuki | Some people say that one can have experiments in natural science, while it is difficult to have meaningful experiments in social science. This must be a big difference. But, lately, aren't people doing game theoretical experiments or experimental economics? |
| Shinzuki | Yes, recently people have conducted quite a few experiments on game theory and economics. But I myself don't understand their aims: They are not really testing theories. To begin with, neither game theory nor economic theory has a testable structure with direct experiments. |

[A waitress brings the food]

**Hammer** Ha, finally the food is here. The fajitas here are quite big. The discussion will have to wait: I want to eat first!

[The three eat quietly for a while]

**Otsuki** Jan, you have a good appetite!

**Shinzuki** He is a healthy man! Even our Genki Morimori can't compete with him.

**Hammer** Is this a compliment? But don't you have high expectations for Genki?

**Shinzuki** Yes, I do. He is studying hard, and more importantly he is keen on discussions. He has been growing a lot. Also, he asks a lot of questions, many of which are elementary and sometimes very basic. Those questions make me rethink seriously such basic issues. He actually has a good influence on me.

**Hammer** Good. I want a graduate student like him. Also, you have the outstanding Toru Majime.

By the way, Oliver, I feel that the crisis of game theory is quite different from the crisis mentioned by Kuhn. Presently as K pointed out, game theory is in a crisis in the sense that there is no substantial development but most of the game theorists don't take the situation seriously.

**Otsuki** Nobody takes the present situation seriously. This is a serious crisis, isn't it? Sorry, sorry, I should be serious.

According to you, game theory seems still to be in the stage of "normal science" and hasn't yet reached the stage of "finding an anomaly".

**Shinzuki** Mm... perhaps, your observation is correct to a certain extent. However, one aspect of game theory may prevent us from finding an anomaly. Namely, game theory has a great power of expression as a mathematical construct. This functions as an obstacle for game theory to develop as an empirical science. Even when it is facing a latent anomaly, game theory is capable of giving an easy resolution by adjusting the structure a little.

**Otsuki** Your opinion sounds interesting, but quite a lot of paradoxes have been discussed particularly in game theory. What do you

|           | think about such paradoxes? Aren't they candidate for anomalies? |
|-----------|---|
| Hammer    | No, I don't think so. For example, the Allais paradox or Ellesberg paradox in expected utility theory didn't succeed in revising the paradigm of utility theory. The revisions proposed are at most slight modifications within the present paradigm. |
| Shinzuki  | I recall the "chain store paradox" of Selten. If you read faithfully Selten's paper, you would find that the paper refutes his famous 1975 paper on perfect equilibrium[2]. Selten himself didn't take a clear-cut stance, but other people took the chain store paradox as an interesting problem to be resolved within the present paradigm of game theory. Then, the field called "refinements of equilibrium" became very popular in the 1980's. |
| Hammer    | Since Allais' and Ellsberg's experiments didn't fit expected utility theory, non-expected utility theory has come about and a huge number of papers have been written. Each paper gives a mathematical generalization of expected utility theory, and claims that the generalized theory is consistent with the experiments of Allais and/or Ellsberg with certain choices of parameter values. It doesn't even reject the original expected utility theory. |
| Otsuki    | Wait, wait! Please be conscious of divergence in your discussions. Each of you is talking about a different subject. Please recall that we have started the discussion on the fact that the great expressing power of game theory may prevent us from appreciating anomalies. What are the points of your arguments on these paradoxes? |
| Shinzuki  | Ah... sorry. I pointed out that these theories have too much freedom for parameters or structures to be adjusted. Conclusions from those theories could be adjustable. Therefore, we |

---

[2] Selten R (1978) Chain store paradox. Theory and Decision 9: 127-159, and Selten R (1975) Reexamination of perfectness concept for equilibrium points in extensive games. International Journal of Game Theory 4: 25-55.

|         | can easily escape from a contradiction with the existing theories. Jan seemed to point out the same. |
|---------|---|
| Hammer  | Indeed, I intended to point it out, too. |
| Otsuki  | Hahaha, I understand both your intentions. Now, you remind me of the Ptolemaic theory of the solar system. The Ptolemaic theory was dominant before the Copernican theory came. As you know, the Ptolemaic theory claimed that the earth was the center of the universe. Although it is now regarded as an ancient silly theory, the theory itself is capable of explaining the movements of the planets. Even if we have more observations, we could explain them by adjusting and generalizing the theory. Then the theory is getting more and more generalized and complicated. After all, however, it is inconsistent with the Newtonian physics in that the Newtonian states that the sun must be the center of the solar system since the mass of the sun is excessively larger than those of the planets. |
| Hammer  | Aha, you compare the present game theory with the Ptolemaic theory of the solar system. It is a nice comparison, indeed. |
| Otsuki  | Nevertheless, a lot of people say now that game theory is the most advanced in social sciences. I'm surprised to hear that the present situation of game theory is so bad. |
| Shinzuki | Unfortunately, I think that is true. I should point out that the situation in economics is pretty bad. But it is different from the critical situation of game theory. Economics often restricts target objects more clearly and discriminates the differences of such objects by various foundational concepts. A lot of pages are spent on discussions of such concepts in good textbooks of economics. |
| Hammer  | Yes, you are right. Textbooks such as *"Principles of Economics"* for first year students are often excellent[3]. Those textbooks discuss carefully the relation between real economic phenomena and theoretical treatments. For example, the distinction of |

---

[3] For example, Mankiw NG (1998) Principles of economics. Harcourt & Company, New York.

|  |  |
|---|---|
|  | the time concept in the theory of firms, "short-run", "long-run" and "super long-run", are very useful distinctions. Nevertheless, such economic contents disappear in textbooks at the intermediate level and many textbooks become simple applications of calculus.<br><br>Macroeconomics may differ from microeconomics, since the connection with real world problems is apparent and a "crisis" might come about. |
| Otsuki | That is an interesting opinion. But since I'm more familiar with game theory than microeconomics and I have almost no idea about macroeconomics, please continue with game theory. |
| Hammer | Okay. Let us continue our discussions on game theory. But, Kurai, why did you address the difference between game theory and economics? |
| Shinzuki | Because I wanted to emphasize the fact that game theory was developed as a mathematical theory from its start. In contrast, economics still has a feature of empirical science. With the lack of the empirical part in game theory, we have very poor vocabulary for evaluating work from the viewpoint of empirical science. Game theorists often justify their work by saying, "the assumption is plausible" or "I'm able to express my intuition in the model". These are no more than subjective statements, and after all, nothing is justified from the empirical point of view. |
| Otsuki | I heard that kind of justification from many game theorists. |
| Shinzuki | Such justifications are not substantive at all. We need certain constraints to prevent such subjective justifications. With no such constraints, we are free to go anywhere, but it would be impossible to encounter serious anomalies. We need suitable constraints on our research. |
| Otsuki | I understand your point. Both of you are very critical about the current state of game theory. From your criticisms, I'm inclined to conclude that game theory can't be expected to have a substantial development as a real science. Since I'm still related to philosophy, I can always return to it. But do you both intend to remain in a field with no future development? |

Shinzuki   Yes, yes, actually I think there is some hope for substantial development in our field. I think, of course, that the field of experimental studies has to develop more. A pure theoretical work is needed for the development of experimental economics itself. Without a theory, one can't make even an experimental design.

Hammer   What kind of hope do you have? Personally I sometimes think it would be better to give up game theory. Do you think it is too late for me to become a singer?

Shinzuki   It is never too late for that. Anyhow, first, one should stop our custom of evaluating results by intuition or mere interpretations. We should consider the foundations of each theory thoroughly. At the same time with the mathematical development of a theory, one should be conscious about what nature and scope the theory has. One has to consider whether the theory is capable of conducting some experiments or has a connection with real world problems. If it is difficult, one has to consider the reason for that difficulty. When experiments or connections with the empirical world are difficult, one should consider why it is difficult and if the theory is still worth something.

I'm not saying that a theory without such a connection is worthless, but instead I want to point out that the present situation without such considerations is problematic. With these, a lot of silly theories could be eliminated.

Otsuki   Are you claiming that philosophical considerations are needed?

Shinzuki   Yes, I am, indeed.

Hammer   That procedure gives a constraint on our research activity, doesn't it? Then we might come across a serious anomalous phenomenon. For example, if contradictions are found between the method and some target objects of a theory even with theoretical and philosophical considerations, then the method itself has to be revised.

Otsuki   I'm surprised to hear that game theory needs philosophical considerations.

Shinzuki  Yes, the role of philosophy is important in the future development of game theory.

Otsuki  Are you finally positive about the future development of game theory?

Shinzuki  Yes, in fact, I think now there is a chance for us to make great progress in game theory. There are plenty of socio-economic problems all over the world. Social sciences including game theory and economics are needed to consider those problems. On the contrary, the framework prepared by Neumann has been exhaustively examined. Now his scope is too narrow to consider present worldwide socio-economic problems. We need a substantial extension of the paradigm. Some people appreciate this situation as a crisis even though it is not in the same sense as Kuhn's. We have reached the stage of a substantial development in game theory as a social science.

A newly built theory may tell that the entire world would have a darker future than ever. Still I think that it is very fortunate as a scientist to participate in or to witness such an academic development.

Otsuki  I thought you were always negative, but now I find you are positive indeed. What kind of concrete research program do you have in your mind?

Shinzuki  For example, I'm thinking of doing research on the internal mental structure of a player. As the description of such a structure becomes richer, we will be able to study how it is formed in society. However, please don't downsize this to a problem of evolutionary development of individual preferences, because then we would remain in the present paradigm.

In particular, I'm working on individual deductive and inductive reasoning in a social situation. Individual reasoning determines individual behavior and then it influences society, and at the same time, the ability for such reasoning is formed in society. For this, I should emphasize that we can analyze the finite or bounded rationality of an individual player.

|           | These will give the foundations of game theory as a positive science, perhaps as a normative science as well, since they will give a lot of hints to consider criteria for personal and social judgments. |
| Otsuki    | I understand what you said. In fact, that is the aim of the present international workshop "Epistemic Logic and Game Theory", isn't it? |
| Hammer    | Even that horrible expression "a landfill of industrial waste" was quite useful to make us think about what is a "crisis" and to consider a possible development of game theory. |
| Shinzuki  | It is "a pile of industrial waste". |
| Hammer    | Okey-doke, whatever you want. Anyway, the aim of the workshop has also become clear. If many questions are asked frankly as in this meeting, then the future of game theory isn't that dark after all.

By the way, why don't we all go to the public bath together after this afternoon session? That public bath really raises the living standard here. Oliver, the public bath itself is fun and it helps you understand Japanese culture too. You should come with us! We will have more naked discussions. |
| Otsuki    | I don't know anything about the Japanese public bath but it sounds like fun. I will go anywhere for fun. |

Narrator: They will go to the public bath. I want to listen to their discussions there, but I can't go with them. I hope they will enjoy their bath and secret discussions.

## Act 4  Decision making and Nash equilibrium

Narrator: A young researcher, Show Hankawa, shows up in this act. He recently returned to Japan after obtaining his Ph.D. in the US. Today, Hankawa comes to Ts University to give a seminar. His seminar starts at 4:30 but he has already arrived at 1:30. He is eager to discuss the latest topics in game theory with Majime. It would be useful for me to listen to what is discussed on game theory abroad. Let's see.

## Scene 1  Recent topics

[Shinzuki and Morimori are in the laboratory and Majime enters with Hankawa]

Majime  Good afternoon, Professor Shinzuki and Morimori. This is Mr. Hankawa, who came from Tokyo today to give a seminar. He received his Ph.D. in the summer of last year at A University in the US, and then he did research in Europe for one year and returned to Tokyo last September. He is working on game theory and industrial organization. He was four years my junior at A University.

Hankawa  It is very nice to meet you. My name is Hankawa.

Majime  This is Professor Shinzuki, and that is Morimori, a graduate student.

Hankawa  Ah... you are THE Professor Shinzuki. I can't believe it is you. I have heard that people expected a lot when you were young. That person is now standing right in front of me. It is a small world. Mr. Majime and I both studied at A University. Where in the US did you study, Professor Shinzuki?

Shinzuki  No, I did my graduate studies at a university in Japan.

Hankawa  I see. That is a pity. We cannot talk about graduate schools in the US.

[Hankawa gives a look at Morimori]

You are Mr. Morimori! Since you are a graduate student, you must have taken the TOEFL test. What is your TOEFL score?

Majime     Hankawa, you are a bit rude. Morimori has no intention in studying in the US. I don't think he has taken the TOEFL test.

Morimori     I'm not good at English. I have never thought about studying in America. But should I?

Majime     I don't think it matters.

Hankawa     However, since the US is the center of game theory, I think it would be more efficient to study there. Of course, the course study is heavy. In every course, the professor gives plenty of reading assignments. If you can live with that, it would be better to study there.

[Hankawa repeats purposely]

    I see, you did *not* even take the TOEFL.

Majime     Well, Hankawa, do you want to discuss anything before your seminar? Or did you come early to prepare?

Hankawa     No, the preparation for the seminar is all done. I didn't come early to discuss anything in particular, but I thought that you might know the latest news from the US. I came early to learn anything new from you.

Majime     I have no particularly interesting news. Also, you have no particular subject for discussion. What shall we do until the seminar starts?

[Morimori says hesitatingly]

Morimori  Before the two of you came in, Professor Shinzuki and I had started talking about how to interpret the Nash equilibrium.

Hankawa  How to interpret the Nash equilibrium? Do you mean Aumann's interactive epistemology approach[1] or are you talking about the evolutionary stability[2]?

Morimori  No, we were talking about the concept of the Nash equilibrium itself.

Shinzuki  I thought I should clarify the relation between the player's decision making and the Nash equilibrium. I brought up this subject, and then Morimori asked how we interpret the Nash equilibrium.

Hankawa  Ah, I see, you are rethinking game theory from the viewpoint of decision theory. One recent issue of *Journal of Theoretical Economics* has a paper where the Nash equilibrium is considered using subjective probability. But now I can't remember the author's name.

Shinzuki  You seem to have read a lot of papers. Actually, only some papers are good, and many others are worthless to read. If a paper has some sign to be categorized into the latter, we could ignore it from the beginning.

Hankawa  Are you telling me I should have ignored the research based on subjective probability?

Shinzuki  As long as you are considering decision making in a game situation with multiple players, it would be better to ignore it. Decision theory is more or less the same as utility theory. Utility theory deals with the individual's taste. Subjective probability is handled as part of utility theory. This means that decision theory treats subjective probability[3] as part of taste. The theory

---

[1] Aumann RJ (1999) Interactive epistemology I and II. International Journal of Game Theory 28: 263-314.
[2] For example, see Weibull JW (1995) Evolutionary game theory. MIT Press, London.
[3] Cf. Savage LJ (1954) The foundations of statistics. John Wiley and Sons, New York.

|  |  |
|---|---|
|  | starts by taking the individual's taste as expressed by a binary relation. Then, it talks only about an axiomatic derivation of a real-valued function representing that binary relation. After all, utility theory is nothing more than a representation theory in the sense that a binary relation between two alternatives is expressed by a real-valued function. Subjective probability is just part of such a real-valued function. |
| Hankawa | What is wrong with it? |
| Shinzuki | The real problem is the structure behind such a binary relation, but not the utility representation. In subjective utility theory, the preference relation is treated as the black box, which is never opened. Utility theory doesn't talk about what subjective probability is or how it emerges in the mind of the individual. Therefore, I don't think that decision theory, or utility theory, is useful for considering decision making in a game situation.<br>What is important is not taste but thought. |

[Shinzuki looks at the three faces in front of him]

| | |
|---|---|
| Majime | I also think that thought but not taste is important for decision making. But something smells in your words. It is my taste, not my thought, that drives me to discuss decision making based on thoughts rather than taste. |
| Morimori | Mr. Majime, wonderful! Let me try some thought experiment, or should I do an experiment in thought? Mm, this is just repetitive. I must be out of thoughts. |
| Hankawa | Wh... what are you doing? I don't understand you. |
| Shinzuki | Sorry for derailing. I was thinking of discussing here decision making in a game situation rather than a one-person decision problem. Then I proposed to discuss about Neumann's minimax theorem. |

[Hankawa, being surprised]

| | |
|---|---|
| Hankawa | Utility theory might be of no use, but you want to discuss Neumann's minimax theorem! That is something from fifty years ago. Over there no one discusses such old tales. Are there still unsolved problems? |

| | |
|---|---|
| Majime | The minimax theorem was proved in 1928 in the famous paper by Neumann[4]. So it is already 75 years old. |
| Morimori | Mr. Hankawa, when you say "over there", do you mean America? In America no one discusses the minimax theorem? Then what do they talk about? |
| Hankawa | Well, the minimax theorem may be briefly mentioned in textbooks but that is only for its historical value, I think. |
| | When I say "over there", of course I mean the US and some universities in Israel and Europe. Is there anyplace else? What is discussed over there is the most recent research done in the top schools in the East Coast, at A University in the West Coast, or in Israel. For example, we discuss what my professor, Scallops, or the famous Ramenski are working on. If you read the papers by those famous people published in journals, you are too late. You need to read the most recent working papers. If you don't do that, you would be far away behind. Mr. Majime must know about the latest working papers and that is why I came earlier. |
| Majime | Indeed, I always give a quick look at the latest working papers, and know the latest scene of the game theory community. But nothing is particularly interesting or important. If you want, I can show you some new working papers from important circles, as well as lists of titles of working papers from all over the world. |
| | But now let's forget about these new working papers or gossip about famous people, and let's discuss today how we should consider the Nash equilibrium. The minimax theorem is also related to the problem of the Nash equilibrium. |
| Shinzuki | That's fine. Please start. |
| Hankawa | Well, that is fine with me too. Listening to old stories once in a while is not too bad. |

---

[4] Von Neumann J (1928) Zur theorie der gesellschaftsspiele. Mathematische Annalen 100: 295-320.

| | |
|---|---|
| Majime | Okay, I will start the discussions on the Nash equilibrium. There are two interpretations of the Nash equilibrium, which I will write down on the blackboard. |

(a): A game is played once, and each player makes his decision before the actual play. A necessary condition for final decisions and predictions is expressed by the Nash equilibrium.

(b): The Nash equilibrium expresses a strategically stable stationary state in a situation where a game has been and will be played repeatedly.

Although the situations are very different in these interpretations, either interpretation will be summarized as the Nash equilibrium in the end. That is, the same equilibrium concept expresses different situations.

| | |
|---|---|
| Shinzuki | Let's call (a) the *ex ante decision interpretation,* and (b) the *stationary state interpretation*. |
| Majime | Indeed, we need names for (a) and (b). Thank you, Sir. Nash's famous paper is written from the viewpoint of (a)[5]. In fact, I think that (a) and (b) can be divided into more detailed cases. For example, the interpretation of the competitive equilibrium in classical economics can be included in (b). Also, the interpretation of the Nash equilibrium in evolutionary game theory should be in (b). According to some people, however, interpretation (b) was also mentioned in Nash's Ph.D. thesis. |
| Hankawa | Do you have two interpretations? Over there, common knowledge is assumed for the Nash equilibrium. |
| Majime | You asked, a minute ago, if we were talking about Aumann's interactive epistemology or about the evolutionary stability. The former is included in (a), and the latter is in (b). |
| Hankawa | In Aumann's interactive epistemology approach or the evolutionary stability, common knowledge is assumed, isn't it? |
| Majime | Oh oh, Hankawa, what did you learn at A University? |

---

[5] Nash JF (1951) Non-cooperative games. Annals of Mathematics 54: 286-295.

Hankawa   Did I say something strange?

Majime    Well, common knowledge is irrelevant in the evolutionary approach. In that approach, each player is identified to be a type of gene, and his behavior is totally decided by that gene. When a new generation comes about, the type more fitting for survival shall leave more of its offspring. Time passes by and when a sufficient number of generations have passed, the distribution of genes converges to the Nash equilibrium. In such an approach, the player isn't a subject that learns knowledge and makes decisions. That is why common knowledge and even individual knowledge are irrelevant in the evolutionary approach.

On the other hand, Aumann's interactive epistemology approach takes the position of (a). Normally in (a) one assumes that the structure of the game is common knowledge. Following Professor Shinzuki, however, this assumption typically remains implicit but is stated as an interpretation.

Hankawa   Is that so? I'm not particularly interested in such interpretations. I have interests in mathematical problems clearly formulated. My goal is to solve unsolved mathematical problems or to generalize theorems that are proven in a special case. I'm interested also in some applications. I want to find some realistic problems and apply game theory to them. However, of course, that is not my true job.

[Majime, getting irritated]

Majime    Hankawa, do you really understand what common knowledge means?

Hankawa   Of course, I know what common knowledge means. It is defined in Aumann's information partition model. I read that paper[6] in a course on game theory at A University. It is not difficult to recall the mathematical definition of common knowledge by looking at the paper.

[Majime turns to Morimori]

---

[6] Aumann RJ (1976) Agreeing to disagree. Annuals of Statistics 4: 1236-1239.

Majime   Morimori, you can say what common knowledge means, can't you?

Morimori   I think I can. A certain statement is common knowledge if and only if the following hold: Every player knows the statement and every player knows that every other player knows and in addition every player knows that every player knows that every player knows, and so on.

Hankawa   Of course, I know it. I thought you asked about a rigorous definition.

Shinzuki   Okay, okay, let's discuss something more concrete. I think we should start with a quick review of the formulation of a game, since Mr. Hankawa might use different terminology. Morimori, could you please write the general definition of the Nash equilibrium on the blackboard?

Morimori   Mm, how general should it be?

Shinzuki   Write the Nash equilibrium for a game with a finite number of players.

[Morimori goes to the blackboard]

Morimori   That is easy. First, we give the definition of a game. Then the Nash equilibrium will be defined.

Definition 4.1: $G = (N, \{S_i\}_{i \in N}, \{g_i\}_{i \in N})$ is a game with $n$ players, where

(1a): $N = \{1,...,n\}$ is the set of *players*;

(1b): $S_i$ is the set of pure *strategies* of player $i$;

(1c): $g_i : S_1 \times ... \times S_n \to R$ is the *payoff function* of player $i$, where $R$ is the set of all real numbers.

Definition 4.2: A strategy combination $s^* = (s_1^*,...,s_n^*)$ is called a *Nash equilibrium* iff for all $i \in N$,

$$g_i(s_i, s_{-i}^*) \leq g_i(s_i^*, s_{-i}^*) \text{ for all } s_i \in S_i, \qquad (4.1)$$

where $s_{-i} = (s_1,...,s_{i-1},s_{i+1},...,s_n)$ and $s = (s_i, s_{-i}) = (s_1,...,s_n)$.

Shinzuki  Could you also write down some simple examples? Mm, what about the Prisoner's Dilemma and the Battle of the Sexes?

Morimori  Sure. On the blackboard the Prisoner's Dilemma is given as Table 4.1. First, the set of players is $N = \{1,2\}$ and the set of strategies is $S_1 = \{s_{11}, s_{12}\}$ and $S_2 = \{s_{21}, s_{22}\}$. The payoff functions $g_1^1$ and $g_2^1$ are given in Table 4.1.

Table 4.1: Prisoner's Dilemma $g^1$

| 1 \ 2 | $s_{21}$ | $s_{22}$ |
|---|---|---|
| $s_{11}$ | 5,5 | 1,6 |
| $s_{12}$ | 6,1 | 3,3 |

Table 4.2: the Battle of the Sexes $g^2$

| 1 \ 2 | $s_{21}$ | $s_{22}$ |
|---|---|---|
| $s_{11}$ | 2,1 | 0,0 |
| $s_{12}$ | 0,0 | 1,2 |

The Nash equilibrium in the Prisoner's Dilemma $g^1 = (g_1^1, g_2^1)$ is given as $(s_{12}, s_{22})$. To see if this pair satisfies Definition 4.2 on the blackboard, we verify the two inequalities given in (4.2).

$$g_1^1(s_{11}, s_{22}) \le g_1^1(s_{12}, s_{22})$$
$$g_2^1(s_{12}, s_{21}) \le g_2^1(s_{12}, s_{22}).$$
(4.2)

In the Battle of the Sexes both $(s_{11}, s_{21})$ and $(s_{12}, s_{22})$ are Nash equilibria.

[Hankawa cannot wait till Morimori finishes his explanation]

Hankawa  You can't prove the existence of a Nash equilibrium within the pure strategies. You need to extend the pure strategies to the mixed strategies. To prove the existence, you use either Brouwer's or Kakutani's fixed-point theorem. When the space of pure strategies is an infinite set, the fixed-point theorem should be extended more. For such an extension, you need knowledge of functional analysis, for example, you start with a Banach space and consider some weak topology. So, you should use a lot of beautiful techniques. These are now quite standard in math-econ[7]. Lately, however, they may not be very popular anymore.

Majime  You are talking about the existence proof of a Nash equilibrium. Could you please write down an example of a game with no Nash equilibrium on the blackboard?

[Hankawa replies in haste]

Hankawa  Well, an example should be one where the payoff functions are not continuous or not quasi-concave. The existence of a Nash equilibrium in a game with discontinuous payoff functions was discussed in a paper in *Journal of Theoretical Economics* a few years ago. If I look at that paper, it shouldn't be difficult to construct an example.

Majime  What? What are you saying?

[Majime looks at Morimori]

Morimori, you can give an example right away, can't you?

Morimori  As a 2-person zero-sum game, the Matching Pennies is such an example, where no Nash equilibrium exists within pure strategies.

Hankawa  I know it, of course.

Majime  Morimori, just to make sure, could you write the Matching Pennies on the blackboard?

Morimori  Sure. It is given as Table 4.3. The two players show their coins simultaneously. If both coins are either head or tails, player 1 wins and he gets the coin of player 2. When the coins show each a different side, then player 2 gets the coin of player 1.

If player 1 in this game changes his strategy in $(s_{11}, s_{22})$ to $s_{12}$, his payoff changes from $-1$ to $1$. Therefore, $(s_{11}, s_{22})$ is not a

---

[7] "Math-econ" is an abbreviation of mathematical economics.

Nash equilibrium. For the same reason the other strategy pairs aren't Nash equilibria either. So this game has no Nash equilibrium within pure strategies.

Since this is a zero-sum game, we can assume that player 2 will minimize the payoff of player 1. Accordingly, it is sufficient to consider only the payoff of player 1.

Table 4.3: Matching Pennies $g^3$

| 1 \ 2 | $s_{21}$ | $s_{22}$ |
|---|---|---|
| $s_{11}$ | 1,−1 | −1,1 |
| $s_{12}$ | −1,1 | 1,−1 |

Hankawa    It is a trivial example. I thought you asked one requiring more mathematical knowledge.

Majime    You seem to be thinking of something very difficult. We will let you explain it if some time is left.

In the Matching Pennies, no Nash equilibrium exists within pure strategies, but if we allow mixed strategies, the existence can be obtained. In that case we use the fixed-point theorem, which seems to be Hankawa's favorite. No, sorry, you like things to be more difficult, don't you, Hankawa?

Shinzuki    Hahaha. Thanks for summarizing. Now, as the preparatory work is done, we can start with the main issue. Well, one often says, "In order for a Nash equilibrium to make sense, the structure of the game needs to be common knowledge". Let's start with this.

Hankawa    But Mr. Shinzuki, isn't that what everybody says? Even at A University, everybody said so. Do we need to discuss something that is already accepted by everybody?

[Majime, irritated]

Majime  I told you, we need to consider if such an opinion is correct. It doesn't matter what everybody at A University says. Here we are trying to discuss something until we all are satisfied. You are incredible! What did they teach you at A University?

Shinzuki  Calm down. Since Mr. Hankawa seems not to be used to this kind of discussions, let's take it easy.

In fact, the minimax theorem for a zero-sum game tells something important for the necessity of common knowledge.

Incidentally, some people claim that the common knowledge assumption is unnecessary for the Nash equilibrium[8]. But the Nash equilibrium itself is a neutral mathematical concept, and is independent of such an assumption. The real question is whether or not a decision criterion or a decision method requires common knowledge.

Hankawa  I still don't see clearly what your problem is, but I will listen just a bit longer.

Shinzuki  Now, let's have some tea.

## Scene 2  Interpretations of the Nash equilibrium

Majime  Before the tea break, we started to consider the necessity of the assumption of common knowledge of the game structure for the Nash equilibrium. Just before the break, Sir, you said that the minimax theorem is relevant for this problem, but I would like to clarify the Nash equilibrium a bit more. Is that okay?

Shinzuki  You are right. First we need to make the problem clear.

Majime  As I wrote on the blackboard, we have two interpretations for the Nash equilibrium. You named (a) the *ex ante* decision interpretation. I heard from some people that this is understood as the self-enforcing property. In a Nash equilibrium, each player has no incentive to change his strategy. Once a Nash equilibrium is reached, the equilibrium itself enforces each player not

---

[8] Cf. Aumann RJ, Brandenburger A (1995) Epistemic conditions for Nash equilibrium. Econometrica 63:1161-1180.

to change his strategy. Thus, the decision of each player stays at the Nash equilibrium. The Nash equilibrium is characterized by the self-enforcing property.

Shinzuki   The argument for the Nash equilibrium in terms of the self-enforcing property was popular in the 1980's. But Majime, you should think about it carefully. The argument says nothing more than a verbal translation of the mathematical definition of the Nash equilibrium. It rephrases Definition 4.2 of the Nash equilibrium in words. The argument can be applied both to the *ex ante* decision interpretation of (a) and to the stationary state interpretation of (b). It only says that if those strategies are chosen, the combination of strategies will be strategically stable. It doesn't say anything about how Nash equilibrium strategies are chosen.

Now, we should consider how to distinguish between the situations behind (a) and (b).

Majime   I see. I'm not able to clarify the problem any further. Sir, please explain the difference between (a) and (b). But I'm afraid that the discussion would digress somewhere if you talk about the two interpretations at the same time. So, first, could you please talk about the *ex ante* decision interpretation (a)?

Shinzuki   No problem. Let me explain (a). Strictly speaking, the problem is not the interpretation of the Nash equilibrium. The problem is the decision making of an individual player from an *ex ante* viewpoint. This problem is closely related to how to interpret the Nash equilibrium. In game theory, the problem of decision making is treated as an interpretation of the Nash equilibrium or other equilibrium concepts. Don't you think that this is clear evidence for the fact that game theory is underdeveloped?

[Hankawa looks surprised]

Hankawa   What? Game theory is underdeveloped? Nowadays game theory is used in so many papers in main economic journals. For example, almost all papers in first-class journals in the US such as *Journal of Theoretical Economics* are related to game theory in one way or the other.

## Scene 2  Interpretations of the Nash equilibrium

Majime  Hankawa, I agree with you that what Professor Shinzuki stated is quite sarcastic. But your argument "game theory is used in so many papers in main economic journals" doesn't imply that game theory is well developed.

Hankawa  But how should I counter-argue Mr. Shinzuki? Game theory and economics are most advanced in the US. It is important to know how game theory is treated over there.

Majime  But it is true that our field has been stagnating worldwide. I think that we should use our own brains to think about the problem.

[Majime looks at Shinzuki]

Did you lead to this digression consciously, Sir?

Shinzuki  No, no, I have no such intention. I might be too hasty with my conclusion. Let me explain my argument a bit more carefully.

First, I look at the present situation of our field. A lot of game theorists take the Nash equilibrium just as for granted, and they are discussing how it should be interpreted. However, our original target is the individual's behavior or decision-making in a game situation. The Nash equilibrium shouldn't be our target, while it plays an important role in the mathematical study of this target.

To be conscious, let me write the target problem on the blackboard. It is:

(c): not "how to interpret the Nash equilibrium"

(d): but how the choice of a strategy is made in a situation with two or more players.

I discuss how and why the Nash equilibrium is relevant to (d).

[Hankawa, shaking his head]

Hankawa  I don't agree with you. A correct interpretation of the Nash equilibrium is given by a mathematical analysis of it.

To begin with, the job of a theorist is to make and refine tools for applied economic research. The consideration of (d) is not a mathematical analysis but is an application. Therefore, it is

Act 4  Decision making and Nash equilibrium   153

sufficient for theorists like us to analyze mathematical properties of the Nash equilibrium. Then, we will have more tools for someone who will use the Nash equilibrium to consider the problem of decision making such as (d).

Pure theory should be done without thinking about applications. Once a pure mathematical research is done, applied people can use it for applied problems including (d).

Morimori   Hm ... I'm not sure that game theory is aimed to provide tools for economic research. But, of course, you are right to say that a pure theory should be pure.

[Morimori, thinking for a while]

But if you make tools without thinking what to use them for, they are of no use and eventually you end up with lots of useless tools. Also, it feels kind of strange to say that pure theory means not to think of applications. Professor, what do you think?

Shinzuki   Theorists often say such a thing. When they are asked about possible applications, they answer that it is theoretically possible to apply the theory to actual problems but, of course, applications will be done in the future by somebody else. In fact, theorists regard such applications as a dirty job, and they often unconsciously believe that pure theory can be applied to actual problems. Applications have never been taken seriously in their minds. And, they claim that it is the duty for pure theorists to make tools and thus to leave applications to those who work on applications.

Hankawa   You speak in a very negative way, but it is truly natural to have a division of labor since our field is already quite large. We need pure theorists and applied people. I'm in the former group. I don't know which you, Mr. Shinzuki, want to be in.

Majime   Mm ... today we have too many digressions from the start. Let's return to our game theoretical problem.

Hankawa   Yes, please.

Majime   Let me summarize what you tried to say, Professor Shinzuki. We should consider the decision making of each player in a

|  | game situation. The Nash equilibrium is a mathematical expression of resulting decisions. Is that correct? |
|---|---|
| Shinzuki | Yes, that is correct. |
| Morimori | Wait a moment. One thing sounds strange. Professor, you often say, "when one discusses a certain property, it is very basic to consider what possesses that property". If I'm not mistaken, you emphasized often in class, "the Nash equilibrium is not a property of a strategy $s_i$ of a player but a property of a strategy combination $s = (s_1, ..., s_n)$ for $n$ players. Mr. Majime said that a Nash equilibrium expresses the decision making of each player. The result of decision making for each player must be one single strategy rather than a strategy combination $s = (s_1, ..., s_n)$. Thus, (d) is a problem for each player, but the Nash equilibrium is the problem for all the players. Is my argument incorrect? |
| Shinzuki | No, no, your argument is an intelligible one. Majime, how do you explain this? |
| Majime | Let me see if I understand his argument. Each player chooses one strategy in his decision making. You wonder what the strategies of the other players $s_{-i} = (s_1, ..., s_{i-1}, s_{i+1}, ..., s_n)$ in the Nash equilibrium are. |
| Morimori | You are right. The problem becomes clearer now. What are they? |
| Shinzuki | That is a quite advanced problem. It must be your problem, Majime. First of all, you should clarify more the situation we are considering. |
| Majime | Yes, let me explain once more our present problem. We are considering what each player will choose from the *ex ante* viewpoint. We suppose that the game is played only once. Before the actual play of the game, each player chooses a strategy to be played. Since each player's payoff is affected by the other players' choices, he needs to consider the decision making of the other players. This part is expressed by $s_{-i}$. |

Shinzuki  Exactly. We should recall that the payoff optimization for player $i$ is not the simple optimization of player $i$'s objective function. For example, in the Matching Pennies of Table 4.3, if player 2 chooses $s_{21}$, the best strategy for player 1 would be $s_{11}$. On the other hand, if player 2 chooses $s_{22}$, the best strategy for player 1 would be $s_{12}$.

Hankawa  Yes, each player needs to think about the other players' choices. This is why common knowledge becomes necessary.

Shinzuki  You are right. Many people say so. But we need to clarify what sneaks into game theory. For example, Mr. Hankawa, you concluded, "this is why common knowledge becomes necessary". Can you explain it?

Hankawa  Well, everybody argues in that way.

Shinzuki  You give a good example of a bad answer. If everybody skips a careful consideration, then the theory is in disaster though everybody follows it.

Okay, please consider first why common knowledge becomes necessary in such a game situation.

Hankawa  Well, one needs to predict the final choices of the others for one's personal decision making, since the others' choices affect one's own payoff. Since the situation is the same for each player, it is possible to predict the choices of the others only when the game is common knowledge.

[Hankawa, showing off]

That is why common knowledge is necessary for decision making in a game situation.

Shinzuki  You should change your last words into "common knowledge is sufficient for decision making in a game situation".

Hankawa  Mm... Yes, you are right.

Morimori  You mean that common knowledge is unnecessary, don't you?

Shinzuki  That is also an abrupt conclusion. For some form of decision making, common knowledge is necessarily involved, but some other form doesn't involve common knowledge at all. To consider the latter possibility, the minimax theorem becomes use-

|  |  |
|---|---|
| | ful. Exactly speaking, the maximin decision criterion, not the minimax theorem, is relevant. |
| Morimori | Professor Shinzuki explained briefly the maximin decision criterion as well as the minimax theorem in our class. Let me recall it. First we take a zero-sum 2-person game. Zero-sum means that the sum of the payoffs for the players is always zero, which is expressed in (4.3): |

$$g_1(s_1,s_2) + g_2(s_1,s_2) = 0 \text{ for all } (s_1,s_2) \in S_1 \times S_2. \quad (4.3)$$

|  |  |
|---|---|
| | The Matching Pennies of Table 4.3 is a zero-sum 2-person game. I remember up to this, but my memory of the maximin decision criterion is vague. |
| Majime | It is easy. I shall explain the maximin decision criterion. |
| | Consider the maximin decision criterion for player 1. When player 1 chooses one strategy $s_1$, his payoff may still vary depending upon player 2's choice. The worst payoff among those possible payoffs is adopted as the evaluation of his strategy $s_1$, and is written as $\min_{s_2} g_1(s_1,s_2)$. Since this worst payoff is regarded as a function of $s_1$, player 1 should maximize this function by controlling $s_1$. This way of choosing a strategy is called the *maximin decision criterion*. |
| Shinzuki | That is correct. The maximin decision criterion suggests player 1 to choose a strategy $s_1$ that maximizes $\min_{s_2} g_1(s_1,s_2)$. This is expressed as $\max_{s_1} \min_{s_2} g_1(s_1,s_2)$. |
| | The maximin decision criterion requires player 1 to think about neither the thought nor the decision criterion of player 2. The faithful reading of the maximin decision criterion is that player 1 should evaluate each $s_1$ by the worst case and based on this evaluation, player 1 should choose the best $s_1$. This is a purely personal thinking, and doesn't involve common knowledge at all. |

Hankawa   Well, but that is the maximin decision criterion, not the Nash equilibrium.

Majime   That's not absolutely right. I think, there is a certain relationship between the maximin decision criterion and the Nash equilibrium. Let me try to recall that relationship.

Well, when we apply the maximin decision criterion to both player 1 and 2, we can write them as

$$\max_{s_1} \min_{s_2} g_1(s_1,s_2) \text{ for } 1$$
$$\max_{s_2} \min_{s_1} g_2(s_1,s_2) \text{ for } 2. \tag{4.4}$$

Then, using the zero-sum conditional (4.3), the criterion for player 2 can be rewritten as $\min_{s_2} \max_{s_1} g_1(s_1,s_2)$.

Morimori   I remember that these two formulae are compared, and that the following inequality holds:

$$\max_{s_1} \min_{s_2} g_1(s_1,s_2) \leq \min_{s_2} \max_{s_1} g_1(s_1,s_2). \tag{4.5}$$

It was easy to show this inequality. But I recall only that there is another theorem about the relationship between (4.5) and the Nash equilibrium. Mr. Majime, what was that?

Majime   Hahaha. I will continue. There is an important theorem which I write on the blackboard:

*Theorem:* Let $g = (g_1,g_2)$ be a zero-sum 2-person game. Then it is a necessary and sufficient condition for (4.5) to hold in equality that $g = (g_1,g_2)$ has a Nash equilibrium[9].

Do you remember this theorem, Morimori and Hankawa?

---

[9] For these results on zero-sum 2-person games, see Luce RD, Raiffa H (1957) Games and decisions. John Wiley and Sons, New York.

Morimori   I remember that theorem. Since the existence of a Nash equilibrium depends upon a game, (4.5) may become an inequality.

Hankawa   I didn't learn about a zero-sum 2-person game, since it is an old subject. But Nash's existence theorem states that there exists a Nash equilibrium.

Majime   You are right. We need to allow the players to use mixed strategies to prove that the game has necessarily a Nash equilibrium. The theorem of the blackboard holds in the case of mixed strategies as well as in that of pure strategies. Therefore, the existence of a Nash equilibrium implies that (4.5) holds in equality. In fact, Neumann proved the existence of a Nash equilibrium in a zero-sum 2-person game within mixed strategies, and thus (4.5) holds always in equality within mixed strategies. This result is called the minimax theorem.

Morimori   Did Neumann prove the existence of the Nash equilibrium within mixed strategies before Nash? Then, why do we use the name Nash for that equilibrium?

Majime   No, exactly speaking, Neumann proved the existence of a saddle point in a zero-sum two-person game within mixed strategies. In a zero-sum 2-person game, the concept of a saddle point is equivalent to the Nash equilibrium. Incidentally, $(s_1^*, s_2^*)$ is called the *saddle point* of $g_1$ when (4.6) holds:

$$g_1(s_1, s_2^*) \leq g_1(s_1^*, s_2^*) \leq g_1(s_1^*, s_2)$$

for all $s_1 \in S_1$ and $s_2 \in S_2$.   (4.6)

Morimori   I remember Neumann's minimax theorem, now. Neumann proved the existence of a saddle point for a zero-sum 2-person game within mixed strategies. Then, (4.5) holds in equality. Therefore, it is called the minimax theorem.

Majime   Then Nash extended Neumann's existence proof of a saddle point to the existence proof of a Nash equilibrium in a game with $n$ players. Of course, his finding relied upon Neumann's

|||
|---|---|
| | theorem. But since Nash's existence proof has had a great influence to economics and game theory, the equilibrium was named after him. |
| Morimori | I didn't know about the background of the Nash equilibrium. |
| Shinzuki | Nash found out his problem since at that time both Neumann and Nash were at Princeton. That kind of academic environment is enviable. |
| Hankawa | Of course, graduate schools in the US are really academic. The movie "Beautiful Mind" depicted Nash's story well, and won an academy award in 2002. When I saw the movie, I felt nostalgic with the atmosphere of top schools. |
| Shinzuki | As a matter of fact, I also went to see that movie. In this spring, I had to give a speech in the orientation for first year students. I wanted to introduce the movie to the students as a story of a great researcher in my field. However, I was very disappointed by the movie. It was a poor movie without any depth. I didn't talk about it in the orientation.<br>The movie said that Nash rewrote economics after Adam Smith. But actually the Nash equilibrium could be introduced because Neumann's minimax theorem existed. The movie didn't describe the relation with Neumann at all. |
| Majime | I didn't see that movie. But no matter how the existence proof of the Nash equilibrium relied upon Neumann's minimax theorem, the Nash equilibrium is a very far-reaching generalization of Neumann's theorem. Thanks to the generalization, game theory and mathematical economics evolved a lot. |
| Shinzuki | I'm not saying that Nash's work is uninteresting but the movie was. Similar to a movie of a few years ago, "Good Will Hunting", it depicts a genius mathematician who solves in a flash mathematical problems lying anywhere. Solving a great problem requires a huge amount of labor and self-sacrifice. Often agony is involved to make others understand how important it is. This isn't described in the movie.<br>Around the 1940's and 50's, a few real monstrous people such as Einstein, Neumann and Gödel worked at Princeton. They |

had reached the highest peaks that mankind had never seen. In such an atmosphere, Nash worked on the Nash equilibrium as an extension of Neumann's minimax theorem. The movie could be slightly deeper if it touched the relation to Neumann and/or the other monstrous people.

Majime  As far as I know, Neumann didn't highly value Nash's existence proof of an equilibrium. He said that the existence proof was more or less the same as Brouwer's fixed-point theorem. But I think that Nash's bargaining theory is very original and beautiful.

Shinzuki  I agree with you. Yet another problem in the movie is that Nash only wants himself to be recognized as a genius and his heart is neither beautiful nor pure. In contrast, Mozart in the movie "Amadeus" is beautiful and pure. The basic difference between "Amadeus" and "Beautiful Mind" is that the director of "Amadeus" loved the main character but the director of "Beautiful Mind" didn't. After all, it is the difference in the abilities between the directors.

Hankawa  It is a problem of taste. I think that "Beautiful Mind" is a very good movie.

Majime  I will have to watch that movie too.

Well, we have discussed the minimax theorem and the existence of the Nash equilibrium.

Shinzuki  Yes, let's return to the problem of game theory. The minimax theorem is about the maximin decision criterion as well as about the Nash equilibrium. Here we emphasize that the evaluation of his strategy is important for the player who follows the maximin decision criterion. He doesn't need to know about the decision criterion of the opponent. Therefore, we don't need to assume common knowledge.

When (4.5) holds in equality, the player's decision making works well. If the opponent player uses the same criterion, equality is reached in (4.5). Then neither player can improve his payoff. In this case, the pair of strategies chosen by the two players becomes a Nash equilibrium.

Act 4   Decision making and Nash equilibrium   161

Morimori   After all, we don't need common knowledge to infer a Nash equilibrium. Are there decision criteria that need common knowledge?

Shinzuki   Yes, there are. When a certain decision criterion is adopted, common knowledge becomes necessary.
By the way, it is almost time for the seminar. Mr. Hankawa, shouldn't you get ready?

Hankawa   Yes, you are right. I better get ready.

Shinzuki   We will have the rest of the discussion tomorrow.

[Hankawa starts preparing for his seminar]

## Scene 3   Infinite regress and common knowledge

[Hankawa comes to the laboratory the next morning]

Hankawa   Good morning, all of you are already here. How are you doing?

Shinzuki   Good morning, I'm fine, thank you, Mr. Hankawa.

Hankawa   By the way, how did you like my seminar yesterday? I wonder if it was okay. Nobody asked questions in the seminar except Mr. Majime.

Shinzuki   Your seminar? Mm ... it was not good. In the beginning of your seminar, Majime asked, "Why do you consider that particular problem?" He didn't ask the particularity or generality of your problem. Instead, he asked its importance. You didn't answer his question directly but immediately started talking about details. Then everybody lost interest. Luckily, Majime helped you out by asking quite a few questions.

Hankawa   I didn't notice that he asked the importance of my problem. But isn't it sufficient to say that it is an unsolved problem?

Majime   You showed the matrix of possible cases on the slide and talked about which cases of the matrix were already studied and which had remained open. You said you wrote a paper because that particular problem was untouched. Your survey of related papers was quite exhaustive, which is useful for people. But you didn't explain the importance of your problem till the end of the seminar.

162   Scene 3   Infinite regress and common knowledge

Hankawa   Is it wrong to solve an untouched problem? This must be the same as "I climb the mountain because it's there".

Shinzuki   Mr. Hankawa, you are quoting the words of Mallory, but you can't use his words to justify your choice of a problem.

George H.L. Mallory was a famous mountaineer, and challenged Mt. Everest three times in the 1920's. He almost accomplished the first ascent of the Everest in 1924 and then he went missing[10]. Even now some people say that the mummy of Mallory is remaining somewhere at the top of the Everest.

In fact, for some purpose, my wife looked for and found the interview with Mallory where his famous answer was given. I think a copy must be somewhere here in the laboratory.

[Shinzuki looks for it in the cabinet]

Here it is. In the interview[11] in 1923, Mallory was asked, *"Why did you want to climb Mount Everest?"* and he answered, *"Because it's there"*.

Then, the interviewer continued, *"But hadn't the expedition valuable scientific results?"* Mallory clearly answered,

---

[10] Edmund Hillary and Norgay Tenzing accomplished the first ascent of the Everest in 1953.

[11] New York Times 1923, March 18.

*"Yes. The first expedition made a geological survey that was very valuable, and both expeditions made observations and collected specimens, both geological and botanical. ... Sometimes science is the excuse for exploration. I think it is rarely the reason.*

*Everest is the highest mountain in the world, and no man has reached its summit. Its existence is a challenge. The answer is instinctive, a part, I suppose, of man's desire to conquer the universe ".*

Now, you understand, Mr. Hankawa, Mallory's intention was truly opposite to your interpretation. For him, the mountain was the Everest, and was the highest and not conquered yet[12]. In fact, there were debates about interpretations of his words in some periodicals for mountaineers in Japan. So, my wife looked for the original interview.

Hankawa    I didn't know such interpretations of his words. But what's wrong with my literal interpretation?

Shinzuki    What's wrong? Would you climb a mountain of industrial waste if it is untouched?

Hankawa    Are you saying that my problem is industrial waste?

Shinzuki    Sorry, that was just an analogy. In our profession, we have so many problems. Most of them are worthless and only a few of them have some value. Before considering a certain problem, you need to discuss its importance.

Incidentally, there are an infinite number of problems, though each is expressed by a finite number of words.

Hankawa    But how do you know which are important and which are not? I think the only way to argue the originality of a problem is to make a literature survey and to show that it is still an untouched problem.

---

[12] Journalist-mountaineer Katsuichi Honda gave a valuable comment on this part.

Majime   Following your argument, we conclude that even climbing a mountain of industrial waste has some value because nobody else has climbed it.

I think, now it is better to return to the subject of yesterday's discussion.

Hankawa   It's fine with me.

Shinzuki   Okay, where shall we start our discussion?

Majime   Yesterday you said, "Common knowledge is inevitable for a certain decision criterion." You are supposed to explain that decision criterion.

Shinzuki   You are right. I shall do that. It is a challenge for me. For simplicity, I limit my argument to a 2-person case.

Now, the argument is as follows[13]: Player 1 predicts the decision of player 2 and chooses a strategy to maximize his payoff based on that prediction. The core of the problem is the prediction by player 1 about player 2's decision. Player 1 assumes that player 2 makes a decision in the same way as himself. Also, player 2 himself makes his prediction and decision in the same way as player 1 does.

Morimori   Professor, it is too abstract to follow.

Shinzuki   Yes, I agree. It is easier to write the argument on the blackboard.

$A$ : Player 1 predicts that player 2 will decide according to $(B)$ below, and then player 1 chooses his best strategy under his prediction based on $(B)$;

$B$ : Player 2 predicts that player 1 will decide according to $(A)$ above, and then player 2 chooses his best strategy under to his prediction based on $(A)$.

---

[13] An introductory but rigorous treatment of the following argument is given in the paper: Kaneko M (2002) Epistemic logics and their game theoretical applications: introduction. Economic Theory 19: 7-62.

Morimori   Wait a moment. Is $(B)$ in $A$ the same as $B$? Similarly, is $(A)$ in $B$ the same as $A$?

Shinzuki   Yes, they are. I put parentheses to $A$ and $B$ inside the sentences $B$ and $A$ to distinguish them.

Morimori   Okay, they can be regarded as the same. Then, your argument sounds very strange. Mm... is it called a *cyclical* argument?

Majime   I think it is cyclical, indeed, since $(B)$ occurs in $A$ and $(A)$ occurs in $B$. Sir, did you write these statements correctly?

Shinzuki   Yes, I think so. This cyclical argument is really the source of common knowledge. Player 1 considers how player 2 makes a decision, and player 2 considers how player 1 makes a decision.

Majime   Can I state your cyclical argument as follows? The sentence $A$ is incomplete since it depends upon $(B)$, and the sentence $B$ is incomplete since it depends upon $(A)$. The completion of $A$ needs $B$, and the completion of $B$ needs $A$.

Shinzuki   Yes, that's right. But you should continue those steps of completion. Graphically they are written as follows:

$$A \to (B \to (A \to (B \to \ldots)\ldots))$$
$$B \to (A \to (B \to (A \to \ldots)\ldots)). \qquad (4.7)$$

The sentences $A$ and $B$ are plugged to $(A)$ and $(B)$ in $B$ and $A$ respectively. Then the resulting sentences need another plugging. The next resulting sentences still have $(A)$ and $(B)$. Hence, we need another plugging again. This process doesn't stop, and we have infinite sequences.

Majime   I think such an infinite sequence is called *infinite regress*.

Shinzuki   You are right. It is called infinite regress. The search for the meaning of each sentence using the other leads to infinite regress.

| | |
|---|---|
| Majime | But in our ordinary thoughts, we regard infinite regress as a difficult paradox or a conundrum. Is it okay to have such infinite regress? |
| Shinzuki | Yes, it is okay. Do you notice that this infinite regress is similar to common knowledge? |
| Majime | Yes, I do. It looks similar to the argument of common knowledge. But I don't see the exact relationship. Please explain it more in detail. |
| Shinzuki | It is better to explain the relationship using belief operators. Let $B_1$ and $B_2$ be the belief operators of players 1 and 2. Here, $B_i(A)$ means that player $i$ ($i = 1,2$) believes $A$. So, $B_i(A \wedge B)$ means that player $i$ believes $A$ and $B$ [14]. |
| | The knowledge is defined to be true belief. Hence, the knowledge of $A$ is expressed as $A \wedge B_i(A)$. |
| Morimori | By $A \wedge B_i(A)$, you mean that $A$ is true and also player $i$ believes $A$. |
| Shinzuki | That's right. In fact, the definition of "knowledge" has a long tradition in European philosophy. Often, it is defined as a verified true belief. This "verified" is quite problematic. To discuss this problem, we need a lot of time. Here, we assume just that "knowledge" is "true belief". |
| Hankawa | It's fine with me, since it is just a problem of philosophy but not of game theory or economics. |
| Shinzuki | Now, using $B_1$ and $B_2$, we can formulate our argument. To begin with, we assume $A \wedge B$. Then player 1 thinks about $(B)$ in the sentence $A$, and player 2 thinks about $(A)$ in the sentence $B$. They believe that their decisions are given by $B$ and $A$. These beliefs are expressed as $B_1(B)$ and $B_2(A)$. Moreover, players 1 and 2 consider $A$ and $B$, respectively. These are $B_1(A)$ and $B_2(B)$. Combining $B_1(A)$ and $B_1(B)$, we have $B_1(A \wedge B)$. Similarly, we have $B_2(A \wedge B)$. |

---

[14] The symbol $\wedge$ means "and". For example, $A \wedge B$ means that both $A$ and $B$ are true.

Majime  Your starting point is $A \wedge B$. Then to complete the meaning of $A \wedge B$, you need to assume $B_1(A \wedge B)$ and $B_2(A \wedge B)$. But now, $B_1(A \wedge B)$ and $B_2(A \wedge B)$ are incomplete. For the completion of these, we assume the next ones, and so on.

Shinzuki  Right. The next step is to add $B_1B_2(A \wedge B)$ and $B_2B_1(A \wedge B)$. This process is described in (4.7). Then, we have all formulae in the following:

$$A \wedge B,\ B_1(A \wedge B),\ B_2(A \wedge B),\ B_1B_2(A \wedge B),$$
$$B_2B_1(A \wedge B),\ B_1B_2B_1(A \wedge B),\ B_2B_1B_2(A \wedge B),\ \ldots \quad (4.8)$$

This means the common knowledge of $A \wedge B$.

Majime  It looks like the common belief of $A \wedge B$. Mm... the first formula $A \wedge B$ is added, and $A \wedge B$ is objectively true. Thus, (4.8) means the common knowledge of $A \wedge B$. Is my understanding correct?

Shinzuki  That is correct.

[Hankawa, being disgusted]

Hankawa  All of you seem very happy to talk about that infinite regress. But if you look at the sentences $A$ and $B$ carefully, they already define a fixed point, and that is a Nash equilibrium.

Shinzuki  That's right. To look at that behavioral consequence from $A$ and $B$, it would better to specify our argument slightly more. Suppose that $s_1$ and $s_2$ are the decision and prediction stated by the sentence $A$, and at the same time that they are the prediction and decision stated by the sentence $B$. Now, $A$ and $B$ are described as $A(s_1,s_2)$ and $B(s_1,s_2)$. Your claim means that $A(s_1,s_2) \wedge B(s_1,s_2)$ implies that $(s_1,s_2)$ is a Nash equilibrium.

Hankawa  That is what I said.

Shinzuki  Anyway, the behavioral consequence from $A$ and $B$ is the Nash equilibrium. However, by extracting the full meaning from $A$ and $B$, we meet the infinite regress of beliefs described by (4.8).

**Morimori**  But (4.8) has an infinite number of formulae, and is hard to imagine it. Is there any more convenient way than (4.8)?

**Shinzuki**  Yes, there is. You can use the *common knowledge logic*, where (4.8) is expressed more concisely[15]. The common knowledge logic has the common *knowledge operator C*. In this logic, (4.8) is expressed as the single formula $C(A \wedge B)$.

**Morimori**  One formula $C(A \wedge B)$ expresses the infinite number of formulae in (4.8). It is much more convenient.

**Shinzuki**  It is more convenient for some purposes, but is less for other purposes.

**Morimori**  It has good and bad aspects as usual. Then, how do you use it?

**Shinzuki**  First, we can prove:

$$\vdash C(A(s_1,s_2) \wedge B(s_1,s_2)) \rightarrow C(Nash(s_1,s_2)). \qquad (4.9)$$

The first symbol $\vdash$ means that what follows $\vdash$ is provable in the common knowledge logic. In words, it is provable that the common knowledge of $A(s_1,s_2) \wedge B(s_1,s_2)$ implies that it is the common knowledge that $(s_1,s_2)$ is a Nash equilibrium. The proof of (4.9) is not difficult. Under a certain condition, the converse holds, too. This converse needs a long argument. So, I will skip it.

**Hankawa**  I would like to hear the proof of (4.9) and its converse. But you can skip them.

**Shinzuki**  I'm sorry about it. But I can tell you about your previous claim. If we focus on behavioral consequences from $A(s_1,s_2)$ and $B(s_1,s_2)$ forgetting their epistemic aspects, then (4.9) becomes:

---

[15] For common knowledge logic, see also the paper mentioned in Footnote 13.

$$\vdash A(s_1,s_2) \wedge B(s_1,s_2) \to Nash(s_1,s_2). \qquad (4.9')$$

But the epistemic aspects involved in $A(s_1,s_2)$ and $B(s_1,s_2)$ are essential. To solve $A(s_1,s_2)$ and $B(s_1,s_2)$ fully, we cannot forget the epistemic aspects.

Our present concern is to discuss not only the derivation of $C(Nash(s_1,s_2))$ from $C(A(s_1,s_2) \wedge B(s_1,s_2))$, but also how it is used. Since, now, $C(Nash(s_1,s_2))$ is our direct target rather than $C(A(s_1,s_2) \wedge B(s_1,s_2))$, I would like to focus on $C(Nash(s_1,s_2))$. Is it fine with you?

Hankawa  I'm fine with focusing on $C(Nash(s_1,s_2))$.

Majime  That's fine with me too. But, Sir, I have one question.

I accept the fact that the decision criterion starting at $A(s_1,s_2)$ and $B(s_1,s_2)$ inevitably involve common knowledge. But does it mean that the game is common knowledge? So far, the common knowledge of a game doesn't appear in your argument.

Shinzuki  You are right, Majime. The discussion until now was about the decision criterion and the derivation of $C(Nash(s_1,s_2))$ from $A(s_1,s_2)$ and $B(s_1,s_2)$. Now, consider the problem of whether the common knowledge of the game is necessary in a concrete game.

Consider a game $g = (g_1, g_2)$ and let $(s_1^*, s_2^*)$ be a Nash equilibrium of $g = (g_1, g_2)$. To obtain $C(Nash(s_1^*, s_2^*))$, it is necessary to assume that the payoff functions of $g = (g_1, g_2)$ is common knowledge.

Hankawa  Is it sufficient rather than necessary?

Shinzuki  Sometimes, you ask a good question. Indeed, I want to say it is not only sufficient but also necessary. The sufficiency part is written as (4.10). Thus, (4.10) means that it is provable that when game $g = (g_1, g_2)$ is common knowledge, it is common knowledge that $(s_1^*, s_2^*)$ is a Nash equilibrium. Since the proof is not difficult, I skip it.

$$\vdash C(g) \to C(Nash(s_1^*, s_2^*)). \qquad (4.10)$$

Hankawa  What happens with its necessity?
Shinzuki  It is much harder to discuss necessity precisely. It is better to skip it.
  [Hankawa, looking triumphantly]
Hankawa  That is fine, but you skip everything.
Shinzuki  I'm sorry.
Morimori  Professor, I have a different question. You assume that various things are common knowledge, which mean that both players think about them in the same way. But, normally, different people think in different ways. One simple example is: the Professor Shinzuki I imagine must be different from the real Professor Shinzuki. Is it okay to assume common knowledge?
Shinzuki  That is exactly the Konnyaku Mondo.
Morimori  But how?
Shinzuki  Let me explain the relation with the Konnyaku Mondo. Now, suppose that player 1 has $C(Nash(s_1^*, s_2^*))$ in his mind. Then, $C(Nash(s_1^*, s_2^*))$ is covered by $B_1$, i.e., $B_1 C(Nash(s_1^*, s_2^*))$. In this case, player 1 believes the common knowledge of $Nash(s_1^*, s_2^*)$, and it may not be objectively true. Then, from the objective point of view, the decision criterion for player 1 is $B_1 C(Nash(s_1^*, s_2^*))$ but not $C(Nash(s_1^*, s_2^*))$. The premise of (4.10) is now $B_1 C(g)$. Hence (4.10) becomes (4.11).

$$\vdash B_1 C(g) \to B_1 C(Nash(s_1^*, s_2^*)). \qquad (4.11)$$

In other words, the derivation of (4.10) occurs in the mind of player 1.

Morimori  Is it related to the Konnyaku Mondo?

Shinzuki  Okay, let me make it a bit clearer. For a moment, suppose that player 1 thinks he is playing the Prisoner's Dilemma $g^1$, and that he believes that $g^1$ is common knowledge between players 1 and 2. However, player 2 thinks they are playing the Battle of the Sexes $g^2$, and he believes that $g^2$ is common knowledge. Notice that the pair $(s_{12}, s_{22})$ is a Nash equilibrium either in $g^1$ or in $g^2$. Thus, each believes that it is common knowledge that $(s_{12}, s_{22})$ is a Nash equilibrium, though each believes actually that a different game is common knowledge among those players. This is expressed in (4.12).

$$\vdash B_1 C(g^1) \wedge B_2 C(g^2) \rightarrow \\ B_1 C(Nash(s_{12}, s_{22})) \wedge B_2 C(Nash(s_{12}, s_{22})). \quad (4.12)$$

Note that in this case, since the game $g^2$ has another equilibrium $(s_{11}, s_{21})$, we have a subtle problem for player 2's decision making. But we would have no problem if we replace $g^2$ by another game where $(s_{12}, s_{22})$ is a unique equilibrium.

Majime  Mm... in (4.12), both misunderstand the situation, and each believes that they play a game, which differs from what the other believes to play. In addition, each believes that the game in his mind is common knowledge. Despite of these misunderstanding, both players derive the same conclusion. I agree with you that this situation is exactly like the Konnyaku Mondo. But where is the real game?

Shinzuki  Indeed, we need to specify a real game. Let's suppose that $g^4 = (g_1^4, g_2^4)$ of Table 4.4 is the real game. We add $g^4$ to the left-hand side of (4.12). Since $(s_{12}, s_{22})$ is a Nash equilibrium in the real $g^4 = (g_1^4, g_2^4)$, the right-hand side of (4.12) can be replaced by the common knowledge of $Nash(s_{12}, s_{22})$. Thus, we have (4.13).

Table 4.4: $g^4 = (g_1^4, g_2^4)$

| 1 \ 2 | $s_{21}$ | $s_{22}$ |
|---|---|---|
| $s_{11}$ | 0,0 | 0,1 |
| $s_{12}$ | 1,0 | 3,3 |

$$\vdash g^4 \wedge B_1 C(g^1) \wedge B_2 C(g^1) \rightarrow C(Nash(s_{12}, s_{22})). \quad (4.13)$$

Majime  The strategy pair $(s_{12}, s_{22})$ is a Nash equilibrium objectively just by chance. It is very similar to the Konnyaku Mondo in that Rokubei actually won from the monk in the debate.

Morimori  Both players think of different games. But the resulting outcome is the same Nash equilibrium as in the real game. In the Konnyaku Mondo, what does correspond to the real game $g^4$?

Shinzuki  Mm ... what should I say? In the Konnyaku Mondo each person is giving his own interpretation to the gestures but to begin with, gestures don't have a real meaning. I see, maybe, the Konnyaku Mondo and (4.13) might be slightly different. Wait, isn't that a general problem of language? The truth doesn't exist separately from people. I have to think about this problem seriously.

Morimori  I will also think about it. Now I have one more question.

Shinzuki  Yes, but please only one question, because I'm already hungry.

Morimori  I think it's simple. According to your explanation, what each player thinks is different from what the other player thinks. Those are even objectively incorrect. Does this argument always involve common knowledge?

Shinzuki   No, I don't think so. I gave the decision criterion $A \wedge B$ as an example where it naturally requires common knowledge. A decision criterion must be one with which a player can make a decision in a coherent manner. For example, player 1 simply assumes that player 2 chooses, say, the first strategy, and 1 maximizes his payoff under this assumption. Player 1 may not think about the other player's decision making.

Morimori   Mm... player 1 doesn't assume that player 2 maximizes 2's payoff. This argument isn't bad, since player 1 can't see 2's mind. Then, a decision criterion doesn't necessarily require payoff maximization, does it?

Shinzuki   That's right. Not necessarily.

Morimori   Then, I can modify your argument to even a simpler one. My criterion is that we should choose the first strategy without thinking about anything. This gives a decision in a very coherent way.

Shinzuki   Haha, you are clever. That is a possible decision criterion. It must be called the *default decision criterion*. It is the categorical imperative for the player to choose his first strategy, forgetting the other people and thinking nothing even about one's own utility. Neither common knowledge nor payoff maximization is needed.

Morimori   If you follow the default decision criterion, you don't need to think about how to behave. It is simple to follow it. I shall use the default decision criterion from now on!

[Hankawa, irritated]

Hankawa   Stupid. It is very stupid. If you take the TOEFL test in that way, you would get only 1/4 or 1/5 of the full score. Then, no graduate schools in the US would approve you. At this university, did you pass exams by that?

Morimori   Yes, sometimes I did, but not always.

Shinzuki   Indeed, if you always use the default decision criterion, people might call you a fool. But isn't it game theoretically clever?

[Hankawa, more irritated]

Hankawa   I have never heard anything like that in our profession. Game theory is supposed to study the behavior of rational people. If we talk only about such stupid behavior, foreign game theorists will call Japanese game theorists "deadwoods".

Shinzuki   You are right. I'm a good or bad example of a "deadwood".

Hankawa   We active game theorists pursue rationality, and we should be rational too. So, we should use high-level mathematics. Yesterday, we discussed the existence theorem of a Nash equilibrium within mixed strategies. We should discuss such high-level mathematics.

Mr. Shinzuki, you seem to like the existence problem, since you quoted somebody's words, "Existence is a challenge". Now I propose to discuss the existence proof of a Nash equilibrium in a more general manner.

Shinzuki   Indeed, Mallory said, "Its existence is a challenge".

Hankawa   Ok, it is a minor correction. Unless existence is proved, we don't know even if each player can choose an equilibrium strategy. So, we should first discuss the existence proof.

Shinzuki   Your problem must be difficult. But I'm already quite hungry. Let's have lunch now. Mr. Hankawa, you still have some time before you return to Tokyo, don't you? In the afternoon you can discuss about the existence of a Nash equilibrium within mixed strategies.

Hankawa   That's fine. I will leave around 3 o'clock, so we still have some time in the afternoon.

[The four leave the stage]

## Scene 4  Mixed strategies

[The four return from lunch]

Hankawa   The cafeteria was very cheap compared to any place in Tokyo. Of course, the quality is so-so and is suitable to the price.

Shinzuki   It's not too bad, is it? Anyway, now you talk about the existence proof of a Nash equilibrium within mixed strategies, don't you?

| | |
|---|---|
| Hankawa | That's correct. It was already mentioned yesterday that Nash himself proved the existence of a Nash equilibrium. I haven't read his original paper, so I don't know the paper itself. But many textbooks mention his proof. First, I shall explain it.<br><br>The case of a finite game is easy and unexciting. In this case, the number of players is finite and each player has a finite number of pure strategies. Within pure strategies, no Nash equilibria may exist like the Matching Pennies as a trivial example, which Mr. Morimori pointed out yesterday. |
| Majime | That's right. Now, you can talk about the existence proof. |
| Hankawa | We extend the set of pure strategies to the set of mixed strategies. When each player uses a mixed strategy, the payoff for him is given as the expected payoff. With this extension, the existence of a Nash equilibrium is proved using Bouwer's or Kakutani's fixed-point theorem[16]. |
| Majime | Up to now, all are written in standard textbooks. |
| Shinzuki | But let me verify something. One mixed strategy is a probability distribution over the set of pure strategies. Each player maximizes the expected payoff. This means that he calculates the probability of each combination of pure strategies before the actual play of the game. |
| Hankawa | That is correct. Is there something you don't understand? |
| Shinzuki | No, I was only checking for my understanding. How does a player actually play a mixed strategy? |
| Hankawa | What do you mean? |
| [Majime, laughing] | |
| Majime | Let me answer it. For example, when player 1 uses the mixed strategy $(2/3, 1/3)$ in the Battle of the Sexes of Table 4.2, he will play a pure strategy $s_{11}$ or $s_{12}$ if it is chosen by the mixed strategy $(2/3, 1/3)$. |
| Shinzuki | I want you to explain a bit more. In the end, each player plays a pure strategy, doesn't he? In the case of the mixed strategy |

---

[16] Cf. Myerson RB (1991) Game theory: analysis of conflict. Harvard University Press, London.

$(2/3, 1/3)$, either $s_{11}$ or $s_{12}$ is finally played with probability $2/3$ or $1/3$. My question is how the probability distribution $(2/3, 1/3)$ is generated.

Hankawa    You ask me how to generate the probability distribution. I don't understand your question.

[Majime, looking askance at Shinzuki]

Majime    You throw a die. When 1 to 4 comes up, you play $s_{11}$, and when 5 or 6 comes up, you play $s_{12}$. Of course, player 1 has to throw the die where player 2 can't see him. Then your next question is how to generate a more complicated distribution such as $(2/13, 11/13)$, Sir.

Shinzuki    That's right.

Hankawa    Should I throw a die with 13 faces? But does such a thing exist?

Majime    I only gave $2/13$ as an example, so it could be $12/23$ or $23/100$.

Hankawa    I see. Thinking about a die is not a good idea.

Majime    For example, you can put 13 balls in an urn, and you write "Win" on two of them. Then you draw one ball out of the urn. For any fraction, we can find a right number of balls to make a generator of any given probability.

Morimori    Do we need 100,000 balls to generate the probability of, say, $12{,}341/100{,}000$?

Majime    Mm ... I say yes, because $12{,}341/100{,}000$ is irreducible.

[Hankawa shrugs his shoulders]

Hankawa    We are interested in a theoretical problem. Who cares about the number of balls needed to play a mixed strategy?

Shinzuki    I care about it.

Hankawa    I didn't ask any question. You took my words literally.

Anyway, when mixed strategies are allowed, we have the existence of a Nash equilibrium. But when the sets of pure strategies are given as continuums and the payoff functions are continuous and quasi-concave, we don't even need mixed strategies for existence. So, we can forget about 100,000 balls. However, when the payoff functions are not quasi-concave or not continuous, we again need mixed strategies to have the existence of an equilibrium. Here, we use high-power mathematics. That is beautiful and excites me a lot.

Morimori  But when the set of pure strategies is a continuum, we have a similar problem with payoffs. For example, when the fractional part of a payoff is $12{,}341/100{,}000$, how do you pay that payoff?

Hankawa  You ask such a terrible question!

Shinzuki  You are right. Sorry, I didn't agree with you, Mr. Hankawa, but I agreed with what Morimori pointed out.

[Hankawa, in a small voice]

Hankawa  Those people are terribly sarcastic.

Shinzuki  Did you say something?

Hankawa  Nothing.

Shinzuki  I see. To begin with, it is an ideal approximation to have any set expressed as a continuum. Of course, a real problem must be finite.

Hankawa  Do you say that a continuum is not real? It is very funny. In any economics textbook, the commodity space is basically a continuum. This commodity space is sometimes approximated with a finite set for some purpose. Nevertheless, reality must be a continuum.

Shinzuki  Since this is an important point, I should explain a bit more. Any goods as well as money have smallest basic units. If we faithfully adopt this observation, the commodity set is discrete or finite. Then you can't use standard mathematical techniques, especially, analysis anymore. Thus we approximate amounts of commodities by continuous variables.

Morimori  According to you, it would be better to assume that all commodities are discrete.

Shinzuki  However, it is fine to make an ideal approximation by a continuum if it is suitable.

| | |
|---|---|
| Morimori | Ah ... it's fine ... if it is suitable. This is a tautology, isn't it? |
| Shinzuki | In fact, it's not so easy to specify when it is suitable. I'll explain it. |
| | Suppose that the basic smallest units are small enough and negligible relative to the amounts in question. For example, the monthly expenditure of money for a household will be more than 1,000 dollars. The smallest unit of money is cent, and so it is less than or equal to $1/100,000$ of the monthly expenditure. This is a very small portion and can be regarded as negligible. In this sense, we can forget the smallest units and can express money amounts by a continuous variable. |
| Majime | Sir, as Hankawa said, in economics, a continuum is typically treated as basic, and a finite case is an approximation. |
| Shinzuki | Certainly, economics and game theory are not well developed to care about such a foundational problem. |
| Hankawa | Again, you reach that conclusion. |
| Shinzuki | Anyway, sometimes it would be better to have an ideal approximation by a continuum, and sometimes it would be better to approximate by discrete variables. For example, the choice of an apartment by a household is better approximated by 0 and 1 variables. But money can be expressed by a continuous variable. |
| Hankawa | You care about when it is approximated by a finite set or when not. In order to forget such a problem, we need a generalization of a model so that it treats both a continuum and a finite set at the same time. |
| Shinzuki | But in such a general model, you obtain only abstract propositions like the existence of an equilibrium or Pareto optimality. I have gotten quite tired of hearing about such results. |
| Hankawa | Regardless of your tiredness, people are working on these problems. |
| Majime | Let's return to our starting point. This morning, Hankawa said, "Unless existence is proved, we don't even know if each player can choose an equilibrium strategy." I think it is reasonable to |

say so. But here, we have a subtle problem. The existence proof of an equilibrium is done by an outside researcher of the game, while the choice of an equilibrium strategy is to be done by an inside player in the game.

Hankawa  That's true, but we assume that the players are in the same position as the researcher. Thus, they prove existence and then choose the strategies constituting an equilibrium.

Majime  That is a standard way of thinking in game theory. It would be better to separate the viewpoint of an inside player from that of the outside researcher.

Hankawa  But when there is a Nash equilibrium, it is sufficient to choose one. I don't see the reason why you need to separate a player in particular from the outside researcher.

Mr. Shinzuki, you claimed that in the case of commodity or money, a continuum is an approximation of a finite case. In the case of probabilities, they are automatically expressed as real numbers. Thus, the set of probabilities is a continuum. If we consider a finite set of probabilities, it must be an approximation of a continuum.

Morimori  Mr. Hankawa, how do you treat a complicated probability such as $12,341/100,000$? Do you prepare $100,000$ or more balls?

Hankawa  You repeat the same question.

Shinzuki  Hahaha, let me write our problems down on the blackboard:

(i) The existence of an equilibrium doesn't necessarily imply that a player can find the equilibrium.

(ii) Even when one actually finds the equilibrium, still there is the problem of generating the probability distribution.

Hankawa  Is it wrong to think that each player is able to find the equilibrium if it exists because he is rational?

Shinzuki  That would be okay only if the term "rational" would include the meaning that if an equilibrium exists, it can be found.

Morimori  That is my favorite tautology.

| | |
|---|---|
| Shinzuki | In such a case, the contents of rationality would be problematic. I don't want to assume by "rationality" that the players are so smart that they can do anything because they are "rational". We should discuss what they can do and what they cannot. |
| Hankawa | What are you thinking of specifically? |
| Shinzuki | I personally think that we should eliminate entirely mixed strategies from game theory, but now I will discuss these problems, assuming that I accept the use of mixed strategies. |
| Hankawa | I personally think that one should use whatever is convenient for writing a paper, but for the moment I will listen to your discussion, assuming that it contains something useful for writing a paper. |
| Shinzuki | Good, okay. First, for (i), we need an algorithm. In a zero-sum 2-person game with mixed strategies, we can find the Nash equilibrium by the simplex method of linear programming. When we can use the four rules of arithmetic $+,-,\times,\div$ and inequality comparisons $\leq$, the Nash equilibrium can be found within a finite number of steps. Accordingly, the player who knows this algorithm can find a Nash equilibrium. This algorithm can be generalized for any 2-person game. |
| Majime | If I'm not mistaken, with the Lemke-Howson algorithm you can find the Nash equilibrium in any 2-person game with mixed strategies[17]. But I have never heard that this algorithm could be extended to cases with more than two players. |
| Hankawa | It must be possible to find such an algorithm even for games with more than two players, because we have already the existence proof of an equilibrium. |
| Shinzuki | But we can easily prove that there is no general algorithm to find the Nash equilibrium in a game with more than two players. |
| Majime | Really? This is the first time I have heard such a thing. |

---

[17] For the Lemke-Howson algorithm as well as treatment by linear programming, see Rosenmüller J (1981) The theory of games and markets. North-Holland, Amsterdam.

Shinzuki  Let's take a 3-person game in which each player has two strategies. The payoffs are given in Table 4.5 and Table 4.6. Depending on the choice of player 3, one of the tables gives the payoffs to the players. For example, when they choose $(s_{11}, s_{21}, s_{32})$, the payoffs are $(2,0,9)$.

Table 4.5: player 3 $s_{31}$

| 1 \ 2 | $s_{21}$ | $s_{22}$ |
|---|---|---|
| $s_{11}$ | 0,0,1 | 1,0,0 |
| $s_{12}$ | 1,1,0 | 2,0,8 |

Table 4.6: player 3 $s_{32}$

| 1 \ 2 | $s_{21}$ | $s_{22}$ |
|---|---|---|
| $s_{11}$ | 2,0,9 | 0,1,1 |
| $s_{12}$ | 0,1,1 | 1,0,0 |

Morimori  What is the equilibrium for this 3-person game?

Shinzuki  This game has a unique Nash equilibrium within mixed strategies. Let $p, q, r$ be the probabilities for the first strategies $s_{11}, s_{21}, s_{31}$ of players 1,2,3, respectively. In the Nash equilibrium, these probabilities are given as

$$p = (30 - 2\sqrt{51})/29, \quad q = (2\sqrt{51} - 6)/21,$$

$$r = (9 - \sqrt{51})/12.$$

(4.14)

Morimori  Professor, how do you calculate that?

Shinzuki  The calculation of (4.14) is tedious but interesting. First, we assume that the players really use mixed strategy $(p, 1-p)$, $(q, 1-q)$ and $(r, 1-r)$ with $0 < p < 1$, $0 < q < 1$ and $0 < r < 1$. In equilibrium, the expected payoffs derived from two pure strategies of a player are the same, provided that the other players use their mixed strategies. Using this fact, we obtain three simultaneous equations of degree 2. We could obtain (4.14) by solving them.

To show uniqueness, we consider the other cases in which pure and mixed strategies are mixed, and show that no equilibria exist in those cases. This step is very tedious.

Hankawa　But it can be calculated. If each player is rational, he will find (4.14) without any difficulty. The real problem is the existence proof.

Shinzuki　You should notice the implication from (4.14) that the Nash equilibrium can't be calculated with any algorithm in the ordinary sense.

Hankawa　But you calculate it as in (4.14), don't you?

Shinzuki　I should explain the meaning of my statement step by step. Recall that the simplex method of linear programming only allows the four rules $+,-,\times,\div$ of arithmetic as well as inequality comparisons $\leq$. Since Tables 4.5 and 4.6 have only integers, we obtain only rational numbers no matter how many times we repeat the four rules of arithmetic. Each probability of (4.14) is an irrational number. Accordingly, there is no algorithm to find the Nash equilibrium within a finite number of steps.

Hankawa　But you, Mr. Shinzuki, solved some equations and obtained (4.14) written on the blackboard. You must have an algorithm.

Shinzuki　In (4.14), I allowed the radical symbol $\sqrt{\phantom{x}}$ besides the four rules $+,-,\times,\div$ of arithmetic. In algebra, the polynomial equation is called *algebraically solved* if a root is obtained by a finite number of applications of the four rules of arithmetic and the radical (root) expressions of any degrees. The Abel-Galois theorem states that some polynomial equation of degree 5 cannot be solved algebraically. However, (4.14) is solvable in this sense[18].

---

[18] Cf. Herstein IN (1964) Topics in algebra. John Wiley and Sons, New York.

Majime     The Nash equilibrium strategies in the 3-person game of Tables 4.5 and 4.6 are described algebraically. Is this true for any 3-person game?

Shinzuki     No, it isn't. When the number of pure strategies in a 3-person game is increased, irrational numbers, much more complicated than those given in a polynomial equation of degree 5, may be involved.

Majime     Does game theory involve such difficult algebraic problems?

Shinzuki     Yes, it does if we really think about equilibrium in mixed strategies. But in principle it is possible to find all the equilibria of finite games. This needs a lot of highly mathematical arguments on the real number system. I don't think it is a good idea to discuss these problems, now.

Instead, I point out two facts. First, Mr. Hankawa, you still seem to think that existence and calculation are equivalent. I show that this is not the case, by proving a certain existence theorem. After that I will address the problem of generating a probability once more.

Hankawa     Finally we arrive at an existence theorem. Are you going to use Kakutani's fixed-point theorem or something else?

Shinzuki     No, I don't use such difficult techniques. Let me write that existence theorem on the blackboard. I will prove it within four lines.

*Theorem:* There are two irrational numbers $a$ and $b$ (maybe, identical) such that $a^b$ is a rational number.

Hankawa     It sounds suspicious. Mm ... but the statement of the Theorem is okay.

Majime     Can you prove it in just four lines? That is something I want to see.

Shinzuki     Let me write the proof on the blackboard.

*Proof.* Consider $\sqrt{2}^{\sqrt{2}}$. There are two cases to be considered.

I: Let $\sqrt{2}^{\sqrt{2}}$ be rational. Then it suffices to put $a = b = \sqrt{2}$.

II: Let $\sqrt{2}^{\sqrt{2}}$ be not rational. We put $a = \sqrt{2}^{\sqrt{2}}$ and $b = \sqrt{2}$. Then $a^b = (\sqrt{2}^{\sqrt{2}})^{\sqrt{2}} = \sqrt{2}^{\sqrt{2} \times \sqrt{2}} = \sqrt{2}^2 = 2$.

| | |
|---|---|
| Majime | Mm, you have proved it indeed. Cases I and II are exclusive and exhaustive. In each case irrational numbers $a$ and $b$ are given. Thus, existence is obtained. |
| Hankawa | You proved it properly. It was not a tricky proof. |
| Majime | But still I think there is something strange. |
| Hankawa | But the existence is proved. |
| Majime | What bothers me is that $a$ differs in cases I and II. Nothing is said about whether I or II is correct. I suppose that II is the case because $\sqrt{2}^{\sqrt{2}}$ must be an irrational number. |
| Morimori | I see, $a = \sqrt{2}$ in I and $a = \sqrt{2}^{\sqrt{2}}$ in II. Then is the existence proved? |
| Shinzuki | That is exactly the problem of this proof. The existence itself is common in either case. However, the proof doesn't specify which the case is.<br>We may have a similar problem with the existence of a Nash equilibrium. The existence proof of a Nash equilibrium might be given, but it is yet another problem if the theorem or the proof gives a specific Nash equilibrium. If a specific Nash equilibrium is not given, the player can't play the equilibrium strategy. |
| Hankawa | But $\sqrt{2}^{\sqrt{2}}$ must be an irrational number, as Mr. Majime said. Therefore, II is the case. |
| Shinzuki | Indeed $\sqrt{2}^{\sqrt{2}}$ is known to be not only an irrational number but also a transcendental number. |
| Morimori | A transcendental number? It sounds very transcendental. |

| | |
|---|---|
| Shinzuki | An algebraic number is a root of some polynomial equation of degree $n$ with rational coefficients. A number is called *transcendental* if it is not an algebraic number[19]. |
| Hankawa | Since any transcendental number is irrational and since $\sqrt{2}^{\sqrt{2}}$ is a transcendental number, $\sqrt{2}^{\sqrt{2}}$ is irrational. Thus, II is the case. |
| Shinzuki | Mr. Hankawa, you used a nice syllogism. However, you should prove the two premises to obtain the conclusion. The major premise that any transcendental number is irrational is immediate, but the minor premise that $\sqrt{2}^{\sqrt{2}}$ is transcendental needed 30 years to be proved. This is related to the 7$^{th}$ problem of Hilbert's 23 Paris problems raised in 1900. The 7$^{th}$ problem asked to show that $2^{\sqrt{2}}$ is a transcendental number. This is equivalent to that $\sqrt{2}^{\sqrt{2}}$ is a transcendental number. The 7$^{th}$ problem was solved affirmatively in around 1930. It took nearly 30 years to prove the 7$^{th}$ problem. |
| Majime | 30 years! Then even if a player is an extremely smart mathematician, a simple existence proof may not immediately give a specific strategy.<br>By the way, who found the proof on the blackboard? |
| Shinzuki | I don't know[20]. It is often used to show the difference between classical mathematics and constructivist mathematics. The proof on the blackboard is accepted in classical mathematics, but is not in constructivist mathematics in that $a$ is not specifically given. Intuitionism is one school of constructive mathematics. Some other time I will discuss the difference between "intuition" used by intuitionists and that by economists or game theorists. |

---

[19] For this theorem, see Anglin WS, Lambek J (1995) The heritage of tales, Section 23. Springer, Berlin.

[20] As far as the author knows, the oldest reference is found in Dummett M (1977) Elements of intuitionism. Clarendon Press, Oxford. In page 9, it is written that the theorem is due to Benenson.

Majime    Can the existence of equilibrium be proved in a constructivist manner within mixed strategies?

Shinzuki    As far as I know, I think so[21]. However, it is quite difficult and often needs tedious steps, since constructivist mathematics prohibits abstract arguments and requires any arguments to be concrete.

Hankawa    I see. Constructivist mathematics seems to be a quite inconvenient world in which to write a paper. But we have a constructive proof of existence. Do we have any further problem?

Shinzuki    You should think about the implication of our argument. Even in mathematics, we have various different rationales. This means that the term "rational" is not uniquely determined at all. When a player is less "rational", we have much more possible meanings. It is not a good idea to say that existence implies calculation if a player is rational. We need to specify what ability he has and what he doesn't.

Morimori    It sounds related to Tolstoi's theorem you mentioned the other day, but I don't recall it.

Shinzuki    It is: *"Happy families are more or less like one another; every unhappy family is unhappy in its own particular way"*[22]. How do you compare this with rationality?

Morimori    I think, *"every very rational thinking is more or less like another; less rational thinking is irrational in its own less rational way"*.

Hankawa    What is that?

Shinzuki    You are right, Morimori. Rationality has a lot of aspects. Perfect rationality means that each aspect is perfect. Thus, perfect rationality is more or less uniquely determined. But "less ra-

---

[21] From the viewpoint of classical logic, this fact is summarized in the working paper: Mere and specific knowledge of the existence of a Nash equilibrium, IPPS-WP No. 741, University of Tsukuba (1997). Readers interested in this paper can contact the author (kaneko@sk.tsukuba.ac.jp). But the paper does not deal with the problem from the constructivist viewpoint.

[22] Act 3, Scene 2.

Act 4 Decision making and Nash equilibrium    187

tional" means that some or many aspects are imperfect. Then the number of such combinations are huge.

Then, shall we discuss another problem?

Hankawa   Another problem? I'm tired, but I can try to be patient a bit longer.

Shinzuki   Let's return to the previous 3-person game given by Tables 4.5 and 4.6. How do you generate the probability $p = (30 - 2\sqrt{51})/29$ of (4.14)?

Hankawa   This time, the probability is an irrational number. I wonder how you generate such a sarcastic problem?

Morimori   Even if a player is rational, he can't generate an irrational probability because he is rational.

Majime     Be serious, Morimori. Since $p$ is irrational, it isn't expressed as a fraction. Accordingly, we can't use the same method as that for $12{,}341/100{,}000$. What should we do?

Shinzuki   In fact, it suffices to have a device to generate the probability $1/10$.

Morimori   That is easy. We prepare an urn with 10 balls with the numbers 0 to 9 written on them and take one ball out of the urn. The probability of taking each ball is $1/10$. But how do you generate the irrational probability?

Shinzuki   We need to repeat drawing a ball from the urn properly. Let me explain. First, let $p = 0.54197...$ be the decimal expansion of $p$. We take the first strategy $s_{11}$ with the following method.

Take one ball from the urn. If we have a ball with a number between 0 and 4, we take strategy $s_{11}$ and if the number is between 6 and 9, we take strategy $s_{12}$. The case of the ball with number 5 remains. In this case, we return the ball to the urn and move on to the 2$^{nd}$ drawing. This process is described in Table 4.7. In the 2$^{nd}$ drawing, the rule is similar to the 1$^{st}$ drawing. If the result is between 0 and 3, we take strategy $s_{11}$, if it is between 5 and 9, take $s_{12}$, and if the result is 4, we move on to

188  Scene 4  Mixed strategies

the 3$^{rd}$ drawing. We continue this process. With this method the probability of strategy $s_{11}$ is $p = 0.54197...$

Table 4.7.

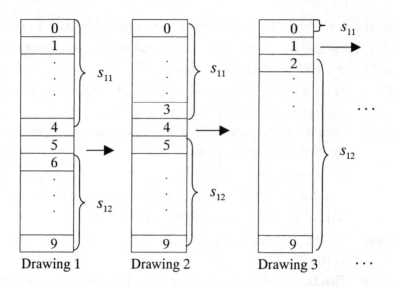

Morimori  Really? Let me calculate it. The probability of having $s_{11}$ in the 1$^{st}$ drawing is $5/10$. The probability of moving on to the 2$^{nd}$ drawing is just $1/10$, and under this condition, the conditional probability of having $s_{11}$ in the 2$^{nd}$ drawing is $4/10$. Looking at this result before the 1$^{st}$ drawing, the probability of having $s_{11}$ in the 2$^{nd}$ drawing is $1/10 \times 4/10 = 4/10^2$. Similarly, the probability of $s_{11}$ in the 3$^{rd}$ drawing is $1/10^3$. Accordingly, the probability of having $s_{11}$ in some drawing is

$$5/10 + 4/10^2 + 1/10^3 + 9/10^4 + ...$$

This infinite sum is exactly $p = 0.54197...$

Majime  As usual, you are good at calculation. The probability of this process reaching the $(n+1)$th drawing is $1/10^n$. Therefore, the probability of this process ending in some finite number of steps is 1. This means that the probability of $s_{12}$ happening is $1-p$. Thus, by this method, player 1 can play the mixed strategy $(p, 1-p)$.

Morimori   Hey, if we use this method, we don't need $100,000$ balls to generate the probability $12,341/100,000$. We have to take a ball from the urn only 5 times to generate $12,341/100,000$, since it is written as $1/10 + 2/10^2 + 3/10^3 + 4/10^4 + 1/10^5$.

Majime   That's right. Morimori, we don't need $100,000$ balls.

Hankawa   I'm pleased with this method, since we can generate any rational and irrational probabilities. Thus there is no difficulty in game theory. Our real problem is to prove the existence of an equilibrium using Brouwer's or Kakutani's fixed-point theorems.

Shinzuki   Mm... I didn't show the above method to claim that you can use mixed strategies.

[Shinzuki, looking at his watch]
   Ah, Mr. Hankawa, it is almost three o'clock.

Hankawa   Yes, you are right. Well, I guess it is time to leave. Today I have learned a lot. For example, I can now rationalize the use of mixed strategies. Any probability is okay when you assume a rational player. It was also nice to hear the fact that the existence of a Nash equilibrium is not a problem from the constructive point of view. I can use this for the justification of my papers I will write from now on.

   Thank you very much for yesterday and today. Soon I will visit again.

[Hankawa leaves the stage quickly]

## Scene 5   Equilibrium as a stationary state

[Morimori is talking to himself in front of the curtain]

We had discussed for such a long time yesterday and today, but we aren't finished yet. These people really like discussions. I was thinking of

## Scene 5 Equilibrium as a stationary state

going home and getting ready for a date with my girlfriend. I'm slightly afraid that my girlfriend will dump me if I become like Professor Shinzuki or Mr. Majime. It is quite understandable why Mr. Majime isn't married yet, though he is trying hard to find a girlfriend. Professor Shinzuki is married but must have a similar problem. His wife may run away from him. I must be careful for myself. So, I should make sure to dress according to the TPO for dating.

[He looks towards the audience]

By the way, did you see the movie "Beautiful Mind"? It has pleased quite a few game theorists and graduate students, giving the hope that game theory will be famous. In fact, I'm happy as well. If game theory becomes famous, it will be easy for me to find a job.

But I don't like those graduate students in the movie. They are competing with each other and are thinking only about their narrow world. Graduate students who are my competitors are all bright and smart like those in the movie. They know a lot but they are boring. I guess Mr. Hankawa was one of them. I wonder if Professor Shinzuki used to be like them. Professor Shinzuki's first name Kurai is almost the evidence for that [23]!

Mr. Hankawa's seminar didn't work well and apparently he had problems with the people here. But, I admire Mr. Hankawa, since he studied in the US, speaks fluent English and is good looking. Moreover, he has already published a paper in *Journal of Theoretical Economics*. Am I inconsistent?

[The curtain opens and Shinzuki and Majime appear on stage]

Shinzuki  Morimori, I heard you were mumbling to yourself. Are you okay?
Morimori  Of course I'm fine. In fact, I have some people to talk to.
Majime  Don't talk nonsense. Well, although we talked a lot yesterday and today, we haven't progressed much. We didn't touch the core of the subjects, maybe because Hankawa gave too many extra comments. Indeed, I was quite surprised about his last

---

[23] One Japanese meaning of "Kurai" is dark.

| | |
|---|---|
| | words. He didn't even try to understand your intention, Sir. Is that the Konnyaku Mondo? Or was it more like Rashomon? |
| Shinzuki | That is his personality. It doesn't change. |
| Majime | We still didn't touch upon the interpretation (b) of a Nash equilibrium as a stable stationary state written on the blackboard. Sir, what do you think about this? |
| Morimori | Hey, do we immediately get back to that kind of conversation? |
| Majime | What else do we discuss? Ah, my words start to sound like Hankawa. |
| Shinzuki | But there isn't much to say about the interpretation of the Nash equilibrium as a stable stationary state. |
| Majime | I think there is something to be discussed. Recently a lot of research has been done on evolutionary game theory. The interpretation of the Nash equilibrium as a stable stationary state is relevant for it. If you don't like talking about evolutionary game theory, I shall explain my ideas. |
| Shinzuki | Okay, go ahead. |
| Majime | I should think about three problems. One is the classical interpretation of equilibrium. You mentioned this interpretation many times[24], and this is important. Due to the game theory literature, I should mention also the repeated game approach as a remark. The last one is the interpretation of the Nash equilibrium in evolutionary game theory. |
| Morimori | Yes, please explain these. |
| Majime | You have repeatedly talked about the perfectly competitive market, in which equilibrium is interpreted as a stationary state. Here, I mention the interpretation of it in the Cournot model. In the Cournot model, the Nash equilibrium is regarded as a stable stationary state in a repeated situation with two or more firms. Since the market is repeated, each firm will know the behavior of the other firms through the past play. Optimization for each firm may become possible once he knows the others' behavior patterns. The other firm will think and behave in the |

---

[24] Cf. Act 3.

|          | same way. A stable stationary state in such a repeated situation is a Nash equilibrium. |
|----------|---|
| Shinzuki | Perhaps, I should give two remarks on that interpretation. There, the Nash equilibrium is only a candidate for such a stable stationary state. It may not be case that the Nash equilibrium will necessarily come about. In a situation with a small number of firms, various behavioral patterns could possibly occur. For example, collusive behavior or other behavior different from the Nash equilibrium may be developed in such a situation. In a situation with many players, it is more certain that the Nash equilibrium will occur as a result. This can be compared to perfect competition or a large city, which we have discussed the other day. |
| Majime   | That's right. What is your other remark? |
| Shinzuki | The other remark is about the relation to the *ex ante* decision interpretation of the Nash equilibrium. After some repetition of the game, each player has accumulated experiences and then he considers the structure of the game based on such experiences. Once he has constructed a view of the game, he can consider his decision from the *ex ante* viewpoint in the model constructed in his mind. Thus, after some repetitions of the game and understanding of the situation, the *ex ante* decision interpretation may become possible. Therefore, the stationary state interpretation and *ex ante* decision interpretation are, in fact, connected. |
| Majime   | Mm... I understand what you said. Also, I understand better why you don't regard the repeated game approach as serious. As you pointed out often[25], the repeated game approach regards the entire repeated situation as a large one-shot game. In this case, we return to the interpretation of a Nash equilibrium as the *ex ante* decision making. This doesn't give a hint to connect those two interpretations. |

---

[25] Cf. Act 1, Scene 4.

Shinzuki  I'm glad you start to understand the relation of the *ex ante* decision interpretation and the stable stationary state interpretation.

Majime  Then, my third problem is the interpretation of the Nash equilibrium in evolutionary game theory. In our profession, many people tend to think that the evolutionary game approach is appropriate for the consideration of repeated situations. What do you think?

Shinzuki  That is another headache, since even the concept of an "individual player" is not specified in the evolutionary game theory, though it is clearer in the original biological evolutionary approach.

In an evolutionary game in biology, a strategy, i.e., a behavior pattern, is identified with a gene. Moreover, an individual player is identified with a strategy or gene. Therefore, a player in one generation always behaves following the same behavioral pattern. The distribution of genes changes over generations through mutations and the law of the survival of the fittest. This implies that one having a higher average payoff will have more descendants. A Nash equilibrium will result after enough time passes by.

Morimori  Why do you have a headache?

Shinzuki  Mm... the description I gave is applied to biology. Evolutionary game theory in economics blindly borrows the formalism. It doesn't talk about what foundational differences are from the standard game theory.

[Shinzuki puts his head in his hands and thinks for a while]

In order to consider human society from a social scientific viewpoint, each theory needs a clear notion of "an individual player". Then, we need to consider the relation between the "individual" and "society". Tomorrow let's discuss about "methodological individualism". Please think about it by tomorrow.

Morimori  Okay, "individualism" ...

Shinzuki  Yesterday and today Mr. Hankawa took part in the discussion. It was different from usual. You learned something, didn't you,

|  |  |
|---|---|
|  | Morimori? In our profession you may see quite a few people like Mr. Hankawa. This was a good exercise for when you will participate in conferences.<br>Well, it is still a bit early but I'm going home. I will see you tomorrow. |
| [Shinzuki leaves the stage] | |
| Morimori | He must be exhausted. I'm really tired. How about you, Mr. Majime? |
| Majime | Of course I'm tired too. But I suppose you learned a lot. |
| Morimori | Yes, thanks to Mr. Hankawa, I now know a bit more about foreign universities. By the way, do you think I should take the TOEFL? |
| Majime | Why not? Of course you have to prepare for it.<br>I work slightly more on my paper now. See you tomorrow. |

Narrator: This act was long and winding. From their discussions, I understand that a lot of complicated problems are involved in game theory. Maybe, targeting human society caused such complications. To begin with, can we really discuss the complications of human society? In the end, Shinzuki said that tomorrow they would discuss about individualism. I have no idea how individualism is related to game theory. Let's wait and see in tomorrow's discussion.

## Act 5  The individual and society

Narrator: As indicated in the previous act, the philosophical foundations of the social sciences will be discussed in this act. Specifically, methodological individualism and its contender, methodological collectivism (holism) will be elaborated on. I'm surprised to learn that there are many isms in our profession and they are fighting each other. I hope that game theory is free from such isms, but if there were some fights, it would be good for progress in game theory. Now, Majime and Morimori start to discuss the problem of what methodological individualism is. Well, let's listen.

### Scene 1  Individualism

[Majime and Morimori are chatting in the laboratory]

Morimori  Yesterday, Professor Shinzuki said that we would discuss individualism today. But I have no idea of what he wants to discuss. He is clearly an individualist, so I don't think he needs to discuss individualism. Do you have any idea what he had in mind for individualism?

Majime  I think he meant methodological individualism.

[Morimori, surprised]

Morimori  Is it a method for individualism? Does it teach how one becomes an individualist? Is it useful for us?

Majime  Please don't ask many questions at once. The term "methodology" points at a scholarly method, and signifies which method should be adopted in the sciences, in particular, social sciences including economics.

Morimori  Okay, it seems more serious than a method to become an individualist.

Majime  Of course, it is a serious problem in social sciences to discuss what method should be adopted to conduct research.

Morimori  But I don't understand yet why scientific research should take some ism, since it is supposed to be neutral.

Majime　　Shall I try to recall what methodological individualism means? It is the ism from an analytical point of view in social science.

Morimori　I'm still bothered by "ism". What does it mean?

Majime　　Ism? I think it points at the research attitude you should take. For example, methodological individualism claims that you should take a viewpoint of methodological individualism in conducting research. The present economics, represented by market equilibrium theory, is constructed from the viewpoint of methodological individualism.

Morimori　You are just repeating "methodological individualism" without giving a further explanation.

Majime　　You are right. I think that it is formulated as, "any social phenomenon can be explained based solely on the characteristics of the individuals".

Morimori　Market equilibrium theory is constructed from that viewpoint, isn't it? The price formation in the market must be a social phenomenon.

Majime　　Yes, I think so.

Morimori　The consumers and producers are the members of the market. Is the price formation reduced to the characteristics of the consumer and the producer?

Majime　　Is anything wrong?

Morimori　I think the numbers of consumers and producers are important factors, too. Without specifying the number, it is impossible to explain the behavior of the market price by analyzing only the characteristics of the individual. This is similar to the fact, Professor Shinzuki pointed out the other day, that the methanol and ethanol are composed of the same atoms, but they have different properties because of their different compositions[1].

Majime　　Mm... you are right. Even in markets with the same characteristics of consumers and producers, the market price may differ when their numbers are different. We can't distinguish between

---

[1] Act 2, Scene 3.

monopolistic and oligopolistic markets, moreover, and perfect competition.

Maybe, something is wrong with the previous formulation of methodological individualism. I recall another statement related to methodological individualism. It was something like "the basic unit for all behavior in society is the individual person".

Morimori   It becomes more philosophical. But how do we interpret that? In market equilibrium theory, a producer is a firm and is a group of individuals. Does it conflict with methodological individualism?

Majime   That's right. Market equilibrium theory conflicts with methodological individualism when group behavior is assumed. Mm... I should admit that I don't understand methodological individualism well.

There is the opposite doctrine to methodological individualism, which is called methodological collectivism or methodological holism.

Morimori   Now you give more isms. But I understand that even you don't understand those isms well. Then, what kind of ism is methodological collectivism?

[Shinzuki appears on the stage]

Majime   Sir, I'm glad you came. Morimori asked me to explain methodological individualism.

I explained that the present economics is based on methodological individualism and that it is formulated as "a social phenomenon can be explained based solely on the characteristics of the individuals". He refuted this by saying that the market price is not determined only by the characteristics of the consumers and producers, but that their numbers are important factors as well.

Shinzuki   What you mentioned is the reductionist version of methodological individual.

Majime   Then I told him another way of defining it, namely, "the basic unit for all behavior in society is the individual person". Then he refuted this again, since a firm is a basic unit in market equilibrium theory but consists of multiple individuals. In either case, market equilibrium theory violates methodological individualism.

Shinzuki   That is the ontological version of methodological individualism.

Morimori   It sounds awfully difficult to understand methodological individualism.

Majime     Just before you came, I referred to methodological collectivism, which should be positioned at the opposite pole of methodological individualism. But this is even harder to understand.

Morimori   Professor, when you mentioned individualism yesterday, did you really mean methodological individualism?

Shinzuki   Yes, I did. I thought I said "methodological individualism".
Majime, you explained both the reductionist version and the ontological version of methodological individualism.

Morimori   I understand the word "reductionist", but not "ontological". What does it mean?

Shinzuki   It means "about the existence of an object" in question. I will explain it more later.

Morimori   Mm... I suppose I understand. Are there still others in addition to reductionist and ontological?

Shinzuki   I would like to add another one. Let me write them down on the blackboard. I will abbreviate them because the full names are too long.

(1): Reductionist individualism
(2): Ontological individualism
(3): Identity-predetermined individualism

[Morimori frowns]

Morimori   I have no idea why we are considering these isms. Could you please first tell me why, and then explain (1), (2) and (3). I'm very curious of what they are.

Shinzuki   I will. Let me start by explaining why we consider these three isms. Various theories have been proposed in game theory and economics. As already discussed yesterday and the day before, even the Nash equilibrium has several interpretations. One theory or one interpretation is constructed to emphasize some selling point. For example, one theory emphasizes its ability to explain certain social phenomena, and another is constructed so that it always gives a solution. Each selling point appears to be nice. Such theories and interpretations have emphases on their own selling points but they often ignore the questions of whether they are sound as a whole. Moreover, once they are constructed, only easy parts tend to be discussed. In some cases, only mathematical structures are paid attention or in some other cases only phenomenal sides are considered.

Morimori  It sounds true. Then, what are you going to discuss?

Shinzuki  My plan is to evaluate those theories and interpretations by comparing and analyzing them. We need some viewpoint for taxonomy in addition to mathematical or phenomenal viewpoints. I adopt methodological individualism and methodological collectivism as key concepts to classify and evaluate extant theories and interpretations. By this taxonomy, important but untouched possibilities may emerge.

Majime  You are going to discuss a meta-theory of various extant theories, aren't you?

Shinzuki  Yes, you could say so. However, my main goal is to separate problems that should be considered but are not really addressed in the extant theories. The following argument will be more or less heuristic, and I won't look for more rigor than what is needed. Sometimes I might use "individualism" or "collectivism" in their daily meanings. Tell me if you don't understand.

Majime  It is true that different methods coexist without being compared in the present economics and game theory. I understand what you are aiming at. So please explain (1), (2) and (3) on the blackboard.

Shinzuki  Okay, I will explain these three isms briefly. I will postpone the detailed discussions about their relations with the present economics and game theory until later.

|  |  |
|---|---|
|  | First, (1): reductionist individualism means that, as Majime said before, a social phenomenon can be explained based solely on the characteristics of individuals. It is named reductionist because the explanations of social phenomena are reduced into the characteristics of individuals. (2): Ontological individualism means that the basic unit for all behavior in society is the individual person, and it doesn't claim any assumptions concerning the characteristics of individuals. Because it emphasizes only the ontological meaning of the basic unit, it is named ontological individualism. |
| Morimori | Now, could you please explain the ontological meaning of the basic unit again? |
| Shinzuki | Mm... it means only who the basic unit is, but not what properties the unit has. Ontological individualism puts an emphasis on the individual unit but is neutral otherwise. In contrast, reductionist individualism goes much further than who the basic unit is. |
|  | Both (1) and (2) were formulated in the 1950s and 1960s through the dispute about methodological individualism[2]. |
| Morimori | Perhaps, I will understand it better if you give some examples. Then what does (3) mean? Every individual is identical? |
| Shinzuki | No, no, the word "identity" means the distinct personality of an individual regarded as a persisting entity. Here I should use the word "subject" rather than "person", since in (3) it may be a firm, company or household etc. Here, the term "identity-predetermined" means that each subject in a theory has already a persisting entity. Thus, (3) means a theory where all the subjects have pre-determined entities. |
| Morimori | Do you allow a firm or company to be a subject in (3)? |
| Shinzuki | Yes. You don't have to consider the individual subject as a human person. But the identity of the individual subject is prede- |

---

[2] A lot of papers were written on this dispute. We refer only to: Lukes S (1973) Individualism, Basil Blackwell, Oxford, and Agassi J (1960) Methodological individualism. British Journal of Sociology 11: 244-270. Related papers are found in their bibliographies.

Act 5  The individual and society    201

termined. In (1) and (2), the individual is taken literally as a human person, as both of you already figured out. Apparently, various theories conflict with (1) and (2). That is why (3) is added.

Morimori　How can we see market equilibrium theory as (3)?

Shinzuki　There, the producer is characterized by a production function and the decision criterion of maximizing profits. When the price is exogenously given, the production function and the decision criterion determine completely the behavior of the producer. Although a firm consists of people, it is regarded as an individual subject in (3).

Also, the consumer is completely characterized by his utility function, income, and the decision criterion of utility maximization. In this sense, the identity of the consumer is also predetermined. Thus, market equilibrium theory can be included in methodological individualism.

Morimori　Will you discuss the relationships among those three? Also, will you discuss how market equilibrium theory differs from game theory from these three viewpoints?

Shinzuki　Yes, I will. In fact, I want to argue that market equilibrium theory is more individualistic than any theory in game theory.

Morimori　I want to hear your argument.

Majime　Wait, wait. I have still several clarifying questions. About the relationship between (1), (2) and (3), am I correct to think that reductionist individualism implies ontological individualism?

Shinzuki　Yes, (1) reduces a social phenomenon into the characteristics of individuals. (1) presumes that the individual is a human. Thus, (1) implies (2). Then, (3) is independent from (1) and (2). Under the assumption that the subject in (3) is a human individual, (1) presumes (3), and (3) presumes (2). Thus, we have $(1) \Rightarrow (3) \Rightarrow (2)$ under that assumption.

Morimori　I don't see the difference between (1) and (3) under that assumption, since the identity of the individual is predetermined both in (1) and (3), isn't it? Then what is really the difference between (1) and (3)?

| | |
|---|---|
| Shinzuki | (1) is assumed to be capable to explain social phenomena by looking solely at the characteristics or identity of the individual. Therefore, the characteristics must be rich enough for such explanations. On the other hand, in (3), for example, the description of a consumer is very poor. It is impossible to explain social phenomena solely by an individual utility function and income. |
| Morimori | Okay, (1) seems stronger than (3). |
| Majime | I understand (1), (2) and (3). You provide those methodological individualisms as keywords, and you will classify and evaluate various extant theories in economics and game theory, in order to look for what is missing. But I still don't see where you really want to go. |
| Shinzuki | Actually, I would like to connect ontological individualism with methodological collectivism. |
| Morimori | It appears again! Could you please explain methodological collectivism, too? |
| Shinzuki | Here, we also have several types. I will write two types down on the blackboard as (4) and (5). |

(4): Ontological collectivism
(5): Individual-forming collectivism

| | |
|---|---|
| Majime | Perhaps, ontological collectivism can be understood similarly to ontological individualism. |
| Morimori | It means that the basic unit for decision making in society is the collective unit, doesn't it? |
| Shinzuki | You are correct. According to a certain dictionary[3], ontological collectivism is described as, |

---

[3] Oxford concise dictionary of sociology. (1994) p.240. Oxford University Press, Oxford.

*"Each social entity (group, institution, society) has a totality that is distinct, and cannot be understood by studying merely its individual elements".*

It is regarded as a claim by the great sociologist Emile Durkheim[4] that *"Social facts can be studied and explained independently of the individual."*

Morimori   It is getting mysterious.

Shinzuki   Sometimes, it becomes more mysterious in the respect that a will is given to each collective entity. I think that this type of thinking has remained from anthropomorphism in the stone ages. As removing "god's will" from science in general, we should remove the will of the whole or the group from social science.

Morimori   I think so too. A "will" is something in the mind of a human person but cannot be found in society. In this case, we better start from the individual human being, don't we?

Shinzuki   Yes, I agree with you. In this sense, I take the position of methodological individualism. However, methodological individualism has also some important aspects to be discussed critically.

[Majime, looking dissatisfied]

Majime   I understand well that the will of the whole or the will of the group is mysterious and is better removed. But if we remove the will, (4) would lose its contents.

Shinzuki   I largely agree with you. But some game theories assume that the whole or the groups have entities as in (4). Therefore, (4) is not completely vacuous as a key concept. We will discuss the details later.

---

[4] Durkheim E (1964, original 1895) The rules of sociological method. Translated by Solovay SA, Mueller JH, Free Press. New York. He is regarded as an exponent of holism. Durkheim himself emphasized social facts with material structures. It is more faithful to interpret his claim as meaning to target social facts that cannot be approached by a reductionist method, than as a holistic claim.

Morimori   But I have no clue to what is meant by (5): individual-forming collectivism.

Shinzuki   We can formulate it as, "each individual is formed in society and society itself is being formed together with individuals". I think that (5) is very important for social sciences.

[Shinzuki, thinking for a while]

Mm, I have a good or bad example to explain the characteristics of (4) and (5). Around 1982, I heard the following example from the famous game theorist, Reinhard Selten. You have heard the name Selten before, haven't you?

Majime   Of course, Selten is famous for the perfect equilibrium point and the chain-store paradox. He received the Nobel Prize in 1994.

Shinzuki   Yes, that is him. In the 1970s, I was strongly influenced by his papers and I wanted to ask him a lot of questions. In 1982, there was a conference at Bielefeld University where he was affiliated, and I went there to participate in the conference. He made some time for me to discuss various themes.

I forgot most of our discussion, except for one conversation about methodological individualism. I still remember clearly his example of holism and his conclusion from it, because I thought that the example and his conclusion should have been reversed, and even now I think so. He mentioned that example, saying "sociologists explain methodological holism in the following way".

Morimori   What kind of example is it?

Shinzuki   Well, I'll explain it to you, imitating Professor Selten.

[Shinzuki, making a severe face and talking extremely authoritative]

"Holism emphasizes that the individual is formed in society and it is nonsense to assume an individual with a perfect identity independent of society. The mushroom nursery is often used as an analogy in order to illustrate this thinking. In the mushroom nursery, there is a seedbed with spores for mushrooms, and in the seedbed the spores are the essential entities. The mushroom buttons that grow above the ground only come

|           | into existence because the seedbed exists. Society corresponds to the seedbed, and the people in society correspond to the mushroom buttons." |
|-----------|---|
| Morimori  | Hey, that is an interesting example. It is true that it is meaningless to separate the individual from society. But this contradicts both (3): identity-predetermined individualism and (1): reductionist individualism. What did you say to Professor Selten? |
| Shinzuki  | I said almost exactly what you said right now, Morimori. "That is an interesting example. It expresses the essence of the relation between the individual and society". But it became strange from here on. Dr. Selten replied in the following manner.<br><br>"No, I gave this analogy as a bad example for holism. We can see by this example that any theory based on holism cannot be an analytical one. The mushroom nursery is a bad example, and holism cannot be analytical." |
| Morimori  | What did you ask next, Professor? |
| Shinzuki  | I asked, "Why is this a bad example? The relation between the individual and society is really similar to that between the mushroom button and the seedbed. Why can't this be analytical?" And I asked him to answer these questions. But all he did was to repeat, "This is a bad example because it is not analytical." He refused to explain any further. This may sound as if he couldn't answer, but his actual talking was so powerful that I was not able to ask any further question. |
| Majime    | Professor Selten meant that science should be analytical. So he said that science was not able to deal with holism, didn't he? |
| Shinzuki  | He didn't answer my questions. So, I didn't understand his true intention. |
| Morimori  | You were treated in such a way a long time ago! |
| Shinzuki  | Of course, quite often. But I also learned from Professor Selten that I should talk in an authoritative way when I say anything negative. |

[Shinzuki, talking in a loud and powerful voice]

              Do you really understand?

| | |
|---|---|
| Morimori | You imitate the strangest things. But I think I will do the same once I have students of my own!<br>But the mushroom nursery is interesting. I would like to hear your opinion about it, Professor. |
| Shinzuki | I can say something about it now. When you interpret the mushroom buttons in the seedbed as mere products, it would be (4): ontological collectivism. The real members of the mushroom nursery are the spores in the seedbed, but the mushroom buttons are mere products of the nursery. In contrast, in the case of human society, society is composed of the individuals, and the individuals are not mere products of society. Nevertheless, the individuals are formed in society and simultaneously they form society. |
| Majime | That seems to be the individual-forming collectivism of (5), doesn't it? |
| Shinzuki | Yes, it is. If we suppose that the essential part of society is not the individual, then the essential elements of society are, for example, institutions or organizations. But it is impossible to take such institutions or organizations as a starting point for research.<br>The mushroom nursery is an important indication for the relation between the individual and the whole, but it is a metaphor. |
| Morimori | I still think the mushroom nursery is interesting. |
| Shinzuki | The interesting point is the indication that the individual is made in society. This feature is missing in game theory and economics, and is only the important point of the mushroom nursery. From now on I will use the mushroom nursery only as a metaphor to indicate that the individual is made in society.<br>Both of you understand what (4): ontological collectivism and (5): individual-forming collectivism are aiming at. |
| Majime | Since (5) regards the individual as made in society, it conflicts with (1): reductionist individualism and (3): identity-predetermined individualism. (2): ontological individualism seems logically consistent with individual-forming collectivism. But it doesn't seem to be connected well either. |

Shinzuki   I think you are right to a good extent. However, I still want to connect individual-forming collectivism with ontological individualism.
           Well, it is lunchtime. Let's continue in the afternoon. I have a meeting at one o'clock. Please come back around three.
[Shinzuki leaves the laboratory]

## Scene 2  Ism

[Shinzuki is coming back to the laboratory, looking happy]
Shinzuki   Finally I have finished all the paperwork. Mm... hey, Morimori, are you taking a nap?
[Morimori gets up from the sofa]
Morimori   No, I was about to get up anyway. Ouah, I'm so sleepy. Discussing for two days in a row made me quite tired. Since we are going to continue, I took a nap to save my energy for more discussion. How was your meeting, Professor?
Shinzuki   The chairman was repeating the same thing over and over, that made me sleepy. But you know, he always speaks in such a loud voice, so I couldn't sleep well. If he was a bit quieter, I could have taken a nap!
Morimori   Hahaha, it was the same here. Mr. Majime was typing constantly on the computer, which bothered me. If he was a bit quieter, I could sleep better.
Majime     This is my job. I have to work. I don't understand how you can sleep in such places. It is rare for me to sleep at the university.
Shinzuki   But that means once in a while you do take a nap! Oh well, time to wake up and get back to work.
           The theme of the afternoon is to use the three versions of individualism and the two versions of collectivism on the blackboard as key concepts in order to compare them with the extant theories and interpretations in economics and game theory. First, let's start with individualism. As I said, the goal is to connect (2): ontological individualism with (5): individual-forming collectivism. We need (3): identity-predetermined in-

dividualism to clarify ontological individualism. But clearly (1): reductionist individualism has little significance. Therefore, we will skip it here.

[Majime interrupts Shinzuki]

Majime     Sir, you have provided the three versions of methodological individualism in order to evaluate extant theories and interpretations in economics and game theory. Are you going to ignore reductionist individualism just by saying, "clearly it has little significance"? When I was at university, I learned that, if I'm not mistaken, methodological individualism is closely related to methodological reductionism in natural sciences. I think it corresponds to (1): reductionist individualism in social sciences. Can't you explain all the isms on the blackboard?

Morimori     I agree with Mr. Majime. So, what does methodological reductionism mean?

[Majime, showing a knowing look]

Majime     In physics, for example, the characteristics and/or structure of materials are divided into molecules, molecules into atoms, atoms into protons or electrons, and these into more elementary particles. The materials are reduced into more basic elements. In this manner, physics has been very successful. This is what is meant by methodological reductionism. Quantum chemistry uses the same reductionist method for chemistry, and biochemistry applies the same to biology.

Sir, you want to skip reductionist individualism, but it follows the legitimate methodology of the present natural sciences. I would appreciate if you explain what is wrong with it.

Shinzuki     Mm, you are not quite right. You claim that that method was successful in physics, chemistry and biology, but your claim is true only for those in which methodological reductionism is successful.

Morimori     It is successful only where it is successful. It is my favorite tautology.

Shinzuki     That's right. I think that reductionist individualism in social sciences is silly, and you would go insane if you think about it

|  | seriously. All I wanted was to evaluate (3): identity-predetermined individualism and to continue quickly to the discussion about (2): ontological individualism and (5): individual-forming collectivism. |
|---|---|
| Majime | But how can you say that reductionist individualism is silly without evaluating it? Is your honorable mind going insane just by thinking about something you don't like? Isn't it the mission for a scientist to explain the reason for his judgment in a way comprehensible to others, not based on what you like or dislike? |
| Morimori | Mr. Majime is right. Professor, stop judging based on what you like or dislike. It was you who said, "It is a matter of thought but not of taste". |

[Shinzuki, against his will]

| Shinzuki | As usual it ends up in a decision by majority. Well, I shall have to start by explaining (1): reductionist individualism. |
|---|---|
|  | As we discussed, there are two kinds of alcohol, namely, methanol and ethanol. Either is composed of carbon, hydrogen and oxygen atoms but because of a different composition of those atoms, methanol is poisonous and ethanol makes you drunk. These properties are determined as chemical characteristics, but aren't determined as the characteristics of the composing atoms. |
| Majime | Actually, Morimori already pointed the same thing. But please stick to reductionist individualism in social sciences. |
| Shinzuki | Okay, okay. Here comes reductionist individualism. But you did start to talk about natural sciences. But fine, I will try not to digress too much. |
| Majime | I admit I talked about natural sciences. But please continue your discussion on social sciences. |
| Shinzuki | Well, reductionist individualism emphasizes that it is sufficient to analyze the characteristics of the human individual for the study of social phenomena. Originally, some sociologists close to psychology emphasized that each social phenomenon can be |

reduced to psychological characteristics of individuals. It is similar to methodological reductionism in physics.

Morimori   Professor, certain social phenomena can be reduced into characteristics of individuals. For example, scissors are made for right-handed people. The reason for it is that most people are right-handed. But, of course, there are also other social phenomena that cannot be reduced to characteristics of individuals.

Shinzuki   Mm... right-handed scissors? Do you know that cheaper ones are usually symmetric?

Morimori   Is that true? Let's look at one.

[Morimori, looking for a pair of scissors in the desk]

Yes, that is right. This is asymmetric one, so it is both-handed.

Shinzuki   That is okay. Consider only expensive and nonsymmetric ones. Suppose that there are right-handed and left-handed scissors. If we try to find out which are used in society, we could pick one individual and examine if he is right-handed or left-handed. In most cases he would be right-handed. We can say that right-handed scissors are used most. In other words, the cause of that can be reduced to the characteristics of individuals.

But think about a hypothetical country where the population consists of 50% right-handed and 50% left-handed people. Then we won't know which type of scissors is used most by observing any one individual. Both types of scissors might be used.

Morimori   In the case of expensive scissors, it must be possible to reduce the phenomenon into the characteristics of individuals.

Shinzuki   I think so. When every individual has the same characteristic, you might be able to reduce the cause to the individual. For example, digestive medicine exists because humans have stomachs and therefore, you can reduce it to a characteristic of the individual. But you don't call this a social problem, do you? To begin with, we call something a social problem when it has some structure irreducible to characteristics of individuals.

|  | However, (1): reductionist individualism has the emphasis that social phenomena are reducible to the characteristics of individuals in general. |
|---|---|
| Morimori | Of course, we can't understand social problems by considering one individual person. If so, I could understand society by thinking about myself. Certainly this is not a natural way of thinking about society. |
|  | I think that an ism causes the present difficulty, since it restricts our way of thinking. We can solve this difficulty simply by forgetting the ism, can't we? |
| Shinzuki | Hahaha, Morimori, I love your flexible mind! Indeed, if we delete the ism, we could solve the difficulty caused by it. |
| [Shinzuki, in thought] | |
|  | However, we can't forget all isms if we seriously conduct scientific research. Sciences and theories have value only if they offer explanations with some universality. We may have interpretations of observed phenomena, but if they were just *ex post* explanations with no universality, they would be useless in that they would offer no predictions in an environment with new factors. Predictions could be possible only if explanations have some universality. To pursue such universality is to have an ism. Therefore, an ism is indispensable for scientific research. Having an ism is the same as pursuing a general principle. |
| Majime | But when we have an ism, people tend to explain everything by that ism. Even methodological reductionism in natural science is problematic for a phenomenon where a collective characteristic has a particular meaning. In the example of alcohol, methodological reductionism claims that we can reduce drunkenness to a characteristic of carbon, hydrogen or oxygen constituting alcohol. I understand what you want to say, Sir. |
|  | On the other hand, if the activities of all basic atoms, carbon, hydrogen and oxygen, are characterized by quantum mechanics, then the chemical characteristics of alcohol are also ultimately explained by quantum mechanics, aren't they? Sir, what do you think about this? |

| | |
|---|---|
| Shinzuki | Using quantum mechanics as the description of basic elements for a chemical theory would not necessarily be methodological reductionism. Even when quantum mechanics are used to describe basic elements, there are still compound characteristics resulting from the combination of those elements. It is the reductionist fallacy to claim that even such compound characteristics can be reduced into properties of more basic elements.<br><br>Social phenomena are compound characteristics. That is why methodological reductionism has no meaning in social sciences. |
| Majime | Sir, you really dislike reductionism. It is better not to ask any further questions about it. |
| Shinzuki | No, it isn't a matter of taste but of thought. |
| Majime | Again! That is already enough of it. |
| Shinzuki | Sorry about it. Anyway, an ism shouldn't be too simplistic. But science would become empty if it has no ism to require any general principle or universality. |

[Morimori, looking confused]

| | |
|---|---|
| Morimori | An ism seems needed, but if it is too strong, it becomes narrow-minded. Where should we draw the line? |
| Majime | After all, we have to evaluate the problem in question carefully, revise the ism within a suitable range, and come up with a more universal form. Ah… this sounds like what you often say, Sir. |
| Shinzuki | Hahaha, now you understand my profound thoughts. |

[Shinzuki, in thought]

An ism must be the research attitude of the researcher. Ideally the ism is reflected in a theory. If a theory is constructed with such an ism and even if the ism is not formulated in a clear-cut manner, it is possible to appreciate the research by moving back and forth between the ism and the theory.

However, practically no one in economics and game theory discusses the ism or principle behind the research. In the first place, a majority of researchers have no ism. Therefore, we change the direction of our way of thinking. Looking back from a theory, we try to find an ism or a principle hidden behind the

|            | theory. Of course, this could be possible only if an ism exists in the theory. Let's try to think about (1) to (5) in this manner. |
|------------|---|
| Majime     | I start understanding your intention. Can I give my opinion on these five isms? |
| Shinzuki   | Yes, please. |
| Majime     | First, (1): reductionist individualism can be seen as a research attitude as well as a characteristic of a theory. However, this ism is too narrow-minded and intolerant. (2): Ontological individualism claims that the decision maker should be a human individual. This is harmless but functions neither as a meaningful characteristic of a theory nor as a research attitude. A meaningful exception is a conflict with the treatment of a firm in market equilibrium theory. (3): Identity-predetermined individualism is also weak as an ism because it seems to be automatically satisfied by any mathematical theory. |
| Morimori   | That is interesting! According to Mr. Majime, (1) is meaningful as an ism but too narrow-minded. (2) has no focus. Any mathematical theory is categorized into (3). Is your classification meaningful, Professor? |
| Shinzuki   | Mm… how can you express such a negative opinion about my classification? But I should ask Majime to continue. |
| Majime     | On the other hand, cooperative game theory is an example of (4): ontological collectivism, and not much else. Cooperative game theory starts with the assumption of cooperative behavior for the players. |
| Shinzuki   | The interpretation of the Nash equilibrium based on common knowledge in non-cooperative game theory is also related to (4). We will discuss this later. |
| Morimori   | But even cooperative game theory considers the incentive of each individual, doesn't it? |
| Shinzuki   | Yes, to some extent. However, the spirit of methodological individualism common under (1), (2) and (3) is that a basic unit of society is an individual. To incorporate firms into market equilibrium theory, (3) deviates from this spirit. |
| Morimori   | That spirit sounds like (2), doesn't it? |

Shinzuki   Yes, therefore, I think that (2): ontological individualism is a pure form of methodological individualism.

Consider cooperative game theory a bit more. We should view cooperation as being formed by the initiation of an individual person. Allowing cooperation of a large group of people is simply against the spirit of methodological individualism. With this violation, except for certain problems, I don't think that cooperative game theory is a serious attempt. Exceptions are, for example, the Shapley-Shubik assignment game and the Nash bargaining game[5]. In both theories, only cooperation of small groups, essentially groups of two people, is allowed.

Majime   But what about the research trend called the *Nash Program* that rethinks cooperative game theory in terms of noncooperative game theory?

Shinzuki   My point is to restrict our attention to cooperative behavior of a few people. The Nash program is to analyze a cooperative game by reformulating it as a noncooperative game. This program itself is neutral with respect to the distinction between a small group and a large group. Therefore, I don't think that the Nash Program has much meaning if cooperative game theory is kept and analyzed in a noncooperative way.

In social sciences, human interactions are basic, but it is quite a cheap attitude to start with groups or the society as a possible unit. Cooperative game theory focuses on human interactions but that formulation is too simplistic. It would be better to forget about cooperative game theory as well as the Nash Program.

Majime   Okay, let's forget cooperative game theory now. I'll return back to my summary.

---

[5] The original paper by Shapley LS, Shubik M (1972) Assignment game 1: the core. International Journal of Game Theory 1: 111-130 is still the best reference dealing with the assignment game. The book by Luce RD, Raiffa H (1957) Games and decisions. John Wiley and Sons, New York is slightly old but detailed on Nash's bargaining theory.

Individual-forming collectivism of (5) sounds appealing, but I don't see its possibilities. First of all, I think it is impossible to make a mathematical model based on this ism. Starting from society as a whole, the composing factors or some parts are determined afterwards. Or should I think that the whole and the parts are determined at the same time? I don't think it is possible to construct a mathematical model reflecting such an aspect. This means that individual-forming collectivism can't be the characteristic of a mathematical theory. To begin with, it can't be a characteristic of a research attitude.

Shinzuki  Are you saying that individual-forming collectivism cannot be a mathematical theory? That might be what Selten meant by repeating, "Holism can't be analytical".

Majime  I think so. Any mathematical method defines the parts and the whole from primitive factors. Therefore, it doesn't make sense to consider a mathematical theory based on individual-forming collectivism.

[Morimori, slightly challenging]

Morimori  I understand your opinion well, Mr. Majime. But, how should we consider the mushroom nursery? The phenomenon that individuals are created in society has not been considered yet. Do you mean that we cannot study such a phenomenon in mathematical theory? I think that is strange.

Majime  Mm ... it is strange to claim that we can't handle a phenomenon like the mushroom nursery with mathematics, because there are no mysterious factors involved.

Shinzuki  Such a mathematical analysis is regarded as impossible, only since nobody has succeeded until now or nobody has seriously tried.

Majime  Sir, you are so optimistic now.

Shinzuki  There should be a mathematical theory that can handle the phenomenon indicated by the mushroom nursery. That is the connection I'm planning to make between (2): ontological individualism and (5): individual-forming collectivism.

Morimori  Could you be a bit more specific, please?

| | |
|---|---|
| Shinzuki | Okay, I mentioned that an ism was a characteristic of a theory and of the attitude of a theorist. However, as for individual-forming collectivism, it would be better to regard it as a characteristic of a research object. In other words, one is aiming at solving the phenomena with such characteristics. Accordingly, I intend to argue that with a mathematical model based on individual-forming collectivism it becomes possible to handle individual-forming collectivism-like phenomena.<br>Mm, I admit that it might also be the characteristic of the research attitude since I say that we should focus on such phenomena. |
| Majime | I understand more or less what you want to say. Some phenomena with the property of (5): individual-forming collectivism do exist indeed. It is the individual's characteristics that come to be formed in society. That is exactly what the mushroom nursery indicates.<br>But I still don't understand how this can be connected with ontological individualism concretely. |
| Shinzuki | Well then, let me move to the discussion of how I will connect these two individualisms. |
| [Morimori interrupts Shinzuki] | |
| Morimori | Professor, you still haven't explained the relation between (1): reductionist individualism, (3): identity-predetermined individualism, and the extant theories in economics and game theory. Could you please start with that explanation? What I actually need is an understanding of the standard theory. |
| Majime | I agree. It will help us understand (5): individual-forming collectivism. |
| Shinzuki | Mm, it seems I have to explain the relationships between (1), (3) and the extant theories in economics and game theory after all. In order to understand those, we need to consider how the relationship between the individual and society is viewed in game theory or market equilibrium theory.<br>But before it, shall we have coffee to wake up? |
| Morimori | I totally agree. How about the coffee shop on the corner? |

[The three leave the stage]

## Scene 3  Connections between the individual and society

[Shinzuki, Majime and Morimori return from the coffee shop]

Morimori   I'm still slightly sleepy. I need stronger coffee than the one we drank in the coffee shop.

Shinzuki   And also a bigger cup!

Morimori   It would be nice to have more coffee shops around here. In Tokyo, there are many cheap coffee chain stores now.

Shinzuki   Indeed, there are many coffee shops in Tokyo. If they would open up a branch next to the university hall, it would make good money. But this town has some cheap sake bars, don't you think? I haven't been for a while. Shall we go for a drink tomorrow?

Majime   Sir, again you talk about alcohol. We should continue the discussion on methodological individualism and collectivism. Before the coffee break, you said we would think about the relation between the individual and society in game theory and economics.

Shinzuki   You are right. We shall think about two extreme cases, through which the other cases will be easier to understand. One extreme case is market equilibrium theory. The other one is the interpretation of the Nash equilibrium based on common knowledge. These two cases are interesting, since the market equilibrium and the Nash equilibrium are close relatives from the mathematical viewpoint, but totally different from the viewpoint of methodological individualism.

Morimori   I know that the competitive equilibrium and the Nash equilibrium are quite similar in their mathematical definitions. Could you please explain their differences with respect to methodological individualism?

Shinzuki   Okay, that is what I want to explain. I will first discuss market equilibrium theory. Then I will talk about the interpretation of the Nash equilibrium based on common knowledge.

| | |
|---|---|
| Majime | I summarize how market equilibrium theory is seen from the viewpoint of methodological individualism. We classify market equilibrium theory into (3): identity-predetermined individualism. The case of pure exchange economy, in which no firms are included, can be classified also into (2): ontological individualism. However, as Morimori already pointed out, it conflicts with (1): reductionist individualism.<br>Is my understanding correct? |
| Shinzuki | Yes, it is. As I emphasized, market equilibrium theory has one radical assumption. That is, market equilibrium theory separates the level of the individual from that of the whole. Of course, it has a connection between the individual and the whole: it is the market price. On the level of the individual, each individual takes the market price for granted, while on the level of the whole, the market price varies depending upon the aggregation of the behavior of the individuals[6]. This separation of the individual from the whole enables us to handle individual behavior in an extremely individualistic manner. |
| Morimori | What do you mean by "extremely individualistic", Professor? |
| Shinzuki | I mean that each individual thinks only about the market price, and optimizes his utility under the budget constraint. Although the market price is the connection between the individual and the whole, this is purely from the objective viewpoint. The individual doesn't take society into account at any rate. Thus it can be called extremely individualistic, can't it? |
| Morimori | So, market equilibrium theory doesn't take society seriously. |
| Majime | Indeed, the market price is the only connection between the individual and society. Thus, market equilibrium theory fails to capture social aspects, doesn't it? |
| Shinzuki | Mm... you tend to take my explanation always so negative. |
| Morimori | But you often end up with a negative conclusion in the end. This time, which is your conclusion, Professor, negative or positive? |

---

[6] Cf. Act 3, particularly, Scene 3.

Shinzuki   In fact, my explanation has both sides. This time, I should start with the positive conclusion.

Market equilibrium theory can be regarded as successful with the radical separation of the individual from the whole. In economic and market phenomena, the connection through the market price is dominant relative to other connections. By that separation, the behavior of the market is more clearly investigated than by taking other minor connections into account. Also from the viewpoint of methodological individualism, it connects the individual nicely to society by the market price, while keeping the treatment of the individual very individualistic. I think that market equilibrium theory is successful not only as a theory of economic phenomena but also as a practical economic institution.

The negative conclusion is more or less what you said: Since it ignores social aspects, the theory can't address more societal problems besides the problems of market phenomena.

Majime   I understand both sides.

Shinzuki   On the other hand, game theory doesn't make such a radical assumption. Game theory inevitably involves some collective components in the connection between the individual and society.

Morimori   Professor, what kind of game theory are you thinking of?

Shinzuki   Any game theory involves some collective elements by and large. For example, consider the case of a zero-sum 2-person game. As I argued before[7], player 1 wants to maximize his payoff function $g_1(s_1, s_2)$ by controlling only his own strategy $s_1$. Player 2 wants to maximize $g_2(s_1, s_2)$, equivalently, to minimize $g_1(s_1, s_2)$, by controlling $s_2$. Player 1 and player 2 affect each other directly through their payoff functions.

Majime   According to the maximin decision criterion, player 1 supposes the worst case for evaluating each of his strategies, because he

---

[7] Cf. Act 4, Scene 3.

doesn't know what player 2 will do. This doesn't mean that player 1 reads the mind of player 2. Nevertheless, player 1 needs to take the possible choices for player 2 into account.

[Morimori, getting excited]

Morimori   I see! I always thought that the maximin decision criterion in a zero-sum 2-person game was quite individualistic. I mean, one's own payoff and the opponent's payoff are exactly opposite and the best thing you can do is to beat your opponent. Therefore, you evaluate your own strategies by the worst scenarios. This is quite individualistic, but still you think about the behavior of the opponent. On the contrary, each individual doesn't think about others at all in market equilibrium theory, though the market involves a lot of people.

Shinzuki   Morimori, I'm impressed. You understand well what we have been talking about!

Morimori   Of course, sometimes I do understand, thanks to you and Mr. Majime.

Shinzuki   Wonderful. I hope you will continue to perform like that!

Now, we can discuss the extreme case in game theory. It is the interpretation of the Nash equilibrium based on common knowledge. Why don't you try to explain that, Morimori?

Morimori   Okay, I will try. As we discussed before, when we consider the Nash equilibrium as the *ex ante* decision making, each player predicts a decision of the opponent by reading the mind of the opponent, and makes his own decision. Accordingly, we move further and further away from individualism.

Majime   That is correct, but Professor Shinzuki asked you to discuss the interpretation of the Nash equilibrium based on common knowledge.

Shinzuki   That's right. Yesterday, we discussed the problem of whether common knowledge about the structure of the game is necessary. Then Majime gave a good summary of the situation.

Majime   I said something like the following: The game $g = (g_1, g_2)$ is common knowledge and both players have exactly the same

|  | decision criteria. They know what they think. All of this becomes common knowledge[8]. In other words, the players share completely the game situation including their thoughts. |
|---|---|
| Shinzuki | Here, the interpretation connects the individual with society by embedding society into the mind of each individual. |
| Majime | It is interesting but is an extreme connection. Then, into which of (1) to (5) on the blackboard do you classify this? |
| Shinzuki | I classify it into both (1): reductionist individualism and (4): ontological collectivism. |

[Morimori, being confused]

| Morimori | Why? They are two opposite poles. Reductionist individualism is the strongest among (1), (2) and (3). Contrarily, ontological collectivism is very collectivistic. Those two must be opposite. |
|---|---|
| Shinzuki | Your thought is natural. However, both poles converge making a full turn, just as the extreme left and the extreme right become similar. As already explained, society exists in the mind of the individual as knowledge. This is collectivistic in that the entire society is in the mind of each individual. Thus, the interpretation of the Nash equilibrium based on common knowledge is categorized into (4): ontological collectivism. |
| Morimori | Then how do you explain that it is classified into reductionist individualism? |
| Shinzuki | Okay, first, recall why common knowledge is needed for the interpretation of the Nash equilibrium. In order for a player to make a decision, he has to picture the situation of the game in his mind. The society as well as the players are in his mind. In this manner, the problem is reduced to the mind of the player. The diagram that I drew on the blackboard yesterday expresses this reduction, which I draw again: |

$$A \to (B \to (A \to (B \to \ldots) \ldots))$$
$$B \to (A \to (B \to (A \to \ldots) \ldots)). \tag{4.7}$$

---

[8] Ibid.

|  | If you want to specify the meaning of $A$, then you need $B$, and if you want to specify $B$, then you need $A$, and so on. The point here is that this argument reduces the problem to one step back. The interpretation of the Nash equilibrium based on common knowledge is reductionist in this sense. When we assume this structure for one player, we can understand the whole problem only by analyzing the thinking of one player. It is individualistic in this sense. Thus, this interpretation can be regarded as one example of (1): reductionist individualism. |
|---|---|
| Majime | Sir, you wanted to skip reductionist individualism by saying that it would drive you insane. However, after all, our majority decision to start with reductionist individualism turned out to be useful. |
|  | But I still have one question. Previously you said that reductionist individualism reduces a social problem to the characteristics of the individual. Now, the problem is reduced into the individual beliefs/knowledge. This is neither a physiological nor a psychological characteristic. Is it okay to include beliefs/knowledge as a characteristic of the individual? |
| Shinzuki | Mm, what are you trying to ask? |
| Majime | I'm asking this because physiological or psychological characteristics are innate traits while beliefs/knowledge are acquired after birth. I thought that characteristics of the individual must be innate traits, and we couldn't call acquired beliefs/knowledge a characteristic of the individual. |
| Shinzuki | Okay, I understand your question. |
|  | Please recall that the distinction between physiological/psychological characteristics and beliefs/knowledge corresponds to that between hardware and software in a computer. A human is also regarded as consisting of hardware and software. Majime, please think about which, the hardware or software of a human being, we should target. |
| Majime | Our hardware has evolved over generations. Also, it is true that one individual has acquired the software, e.g., beliefs/knowl- |

|  |  |
|---|---|
| | edge and behavioral criteria, by living in society. So, we should target software. |
| Shinzuki | Then, what do we, social scientists, analyze in terms of reductionist individualism? |
| Majime | Perhaps, we analyze social phenomena by it. Ah, I see, if we try to explain social phenomena by reducing social phenomena to the hardware of the individual, then we could explain social phenomena by the evolution of our hardware such as physiological/psychological elements. This must be funny. You want to lead us to the conclusion that the software is included in characteristics of the individual. |
| Shinzuki | I don't deny the possibility that some social phenomena are related to our physiological elements. Nevertheless, the software of the individual is really our target of even (1): reductionist individualism. |
| Morimori | But you want to discuss (5): individual-forming collectivism. Please discuss it. |
| Shinzuki | That's right. But today, we don't have much time left to discuss it. I suggest we will do it tomorrow. |
| Majime | It is okay with me. But may I summarize what I understand from your discussion on (1) to (5)? It won't take long. |
| Shinzuki | I still have some time before shopping. So go ahead. |
| Majime | Market equilibrium theory is very individualistic in that the individual has to think only about his behavior. On the other hand, the reductionist interpretation of the Nash equilibrium embeds the whole society into the mind of the individual. This is very collectivistic. These are extreme cases. I still think that there should be some natural relation between the individual and society, unlike the two extreme cases. It must be the key for (5). To be honest, however, I still don't see the connection between (2): ontological individualism and (5): individual-forming collectivism. |

[Morimori, shaking his head]

| | |
|---|---|
| Morimori | No, no! I feel something strange. Both of you say that market equilibrium theory is extremely individualistic. I followed your |

|         | explanation. But in Mr. Majime's graduate class, we discussed the Debreu-Scarf limit theorem on the core and competitive equilibrium[9]. And in your undergraduate class, Professor, you used the Edgeworth box diagram to explain the convergence of the core to the competitive equilibrium. Both you and Mr. Majime said that this theorem explains *the law of indifference,* i.e., the market allows one price for one good. But the core is a concept in cooperative game theory, and we assume arbitrarily large coalitions for this theory. In other words, the cooperation of so many agents is allowed. Isn't it quite collectivistic? |
|---|---|
| Majime | That is right. I also felt that there was something missing in our discussion. In order to prove the limit theorem, we need to assume coalitions of extremely large numbers of agents. Morimori is right in that collectivistic elements are hiding in market equilibrium theory. Am I correct, Sir? |
| Shinzuki | No, you are not completely correct. In fact, a certain mathematical structure that was introduced for a different purpose causes this mischief. Actually, I don't like discussing this problem, since it is a delicate problem, and I don't want to judge the Debreu-Scarf theorem. I think that their explanation of the price formation is a great achievement. I still clearly remember that I was deeply impressed when I read their paper. So, I don't like to convict their work, but now it appears I can't avoid it. |
| Majime | It seems you are turning into Oedipus again. |
| Morimori | It has been a while! |
| Shinzuki | Yesterday, we discussed about the probabilities represented by irrational numbers in the existence proof of a Nash equilibrium[10]. We also discussed the assumption of the commodity space to be a continuum. This ideal approximation is very convenient for an analytical purpose. However, this allows the |

---

[9] Debreu G, Scarf H (1963) A limit theorem on the core of an economy. International Economic Review 4: 235-246.

[10] Act 4, Scene 4.

|          | quantities of irrational numbers to sneak into the commodity space. |
|----------|---|
| Majime   | Indeed, a competitive equilibrium may involve a quantity represented by an irrational number in the standard theory. But is it related to the Debreu-Scarf limit theorem? |
| Shinzuki | Yes, it is, indeed. We need an extremely large coalition to deal with resulting quantities in trade expressed by irrational numbers. When an irrational number is approximated by a rational number, the denominator and numerator of that rational number will be large integers. Large coalitions are needed to express those large integers. If we make better approximations, those coalitions are getting larger and larger. For this purpose arbitrarily large coalitions are needed. Thus, the continuum of the commodity space causes the need for large coalitions. |
| Majime   | I shall check the proof of the Debreu-Scarf theorem once more. Sir, are you saying that such a problem arises because we assume perfect divisibility for each commodity? |
| Shinzuki | Yes, I am. The most successful treatment of indivisible goods is the Shapley-Shubik assignment game. In that game, the core and the competitive equilibria coincide with each other without assuming large coalitions[11]. We need only any pair consisting of a buyer and a seller. In such a case, a collectivistic problem of huge coalitions doesn't arise. Therefore, the law of indifference can be explained with the equivalence of the core and the competitive equilibria in the assignment game. |
| Morimori | I'll have to read the paper by Shapley and Shubik, too. There are so many papers I have to read. |
| Shinzuki | But we have a quite limited number of different ways of thinking. Once you know them, you can skim through the rest and look for new ideas only. |
| Majime   | Sir, after reading many papers, you can say so, but before you understand, it feels as if there are an infinite number of papers to be read. |

---

[11] See Footnote 5.

Morimori   What should I say?
Shinzuki   You don't need to say anything and just do it.
           Today was a quite long discussion. And we haven't yet touched the most important part, namely (5): individual-forming collectivism. Let's continue tomorrow morning. Tomorrow I have no meetings and no lectures. We can devote the whole day to the discussion. Well, it is about time for me to leave for shopping. I'm looking forward to seeing you tomorrow.
[Shinzuki leaves the stage]
Morimori   We have been discussing things for three days continuously. I'm tired. Though it is still bright, I will go home and straight to bed.
Majime     Really? I should continue writing on my paper. See you tomorrow.
[Morimori leaves the stage. Majime gets back to work on his computer]

## Scene 4   Internal mental structure of the individual

[Shinzuki and Majime are sitting when Morimori appears on the stage]
Morimori   Good morning. Did I keep you waiting?
Shinzuki   Yes, just about 30 minutes.
Morimori   I'm sorry. I was so tired yesterday. So I went jogging and then straight to bed. But I didn't wake up this morning.
Shinzuki   You are young! Also, it is good to hear that you are working out!
[Majime has a dissatisfied look on his face]
Majime     I'm still young too but I am punctual. Morimori eats well, sleeps well and speaks a lot. He is a quite inefficient person. But it is good that he is energetic when he is awake.
Morimori   You are very efficient in whatever you do, Mr. Majime. But life isn't interesting if everything is efficient. Research should be like life; we should eat well, sleep well and work with pleasure.
Shinzuki   Hahaha. By the way, today we discuss (5): individual-forming collectivism, and its relation to (2): ontological individualism. For this, I'll start by talking about the internal structure of the

|  | individual. I shall argue about what internal structure of the individual is assumed in market equilibrium theory or game theory. |
|---|---|
| Morimori | By the internal structure of the individual, do you mean, first the mouth, then the gullet, the stomach, after that the duodenum, the small intestines, the large intestines, the rectum and finally the anus? |
| Majime | Yes, certainly, animals have a structure like a pipe. Food enters through the mouth, part of it is absorbed but most of it is excreted via the anus. Incidentally, the larva of a dragonfly has a structure like a jet engine. She moves forward by drinking water through her mouth and gushing it out through her back. Humans are also input-output machines with things entering the mouth and leaving through the anus. |

[Majime, slightly embarrassed by himself]

|  | Ah... Morimori succeeded in taking me in. |
|---|---|
| Shinzuki | Indeed, it is unlike you, Majime. But what you said is related to our discussion. Recall that in individual-forming collectivism, we consider how the individual is formed in society. That is why I plan to think about the internal structure of the individual. |
| Majime | I'm sorry. In individual-forming collectivism, the identity of the individual is formed in society, similar to the metaphor of the mushroom nursery. But organs are not formed in society and are determined as a result of thousands of generations of evolution. |
| Shinzuki | We can regard a human as consisting of hardware and software like a computer. Hardware like the organs can be seen as constant for one generation, but the part of the individual that is formed in society must be software.<br><br>Perhaps, I should use "the internal *mental* structure of the individual" rather than simply "the internal structure". |
| Morimori | Then, we don't misunderstand. |
| Shinzuki | However, before we turn our attention to the internal mental structure itself, I would like to consider how the extant theories |

|  |  |
|---|---|
|  | treat it. Let's start with the internal mental structure of the individual in market equilibrium theory and game theory. |
| Majime | Mm, the problem has hardly been addressed. In market equilibrium theory, the consumer consists of a utility function and income, or maybe, the labor productivity. Only the utility function is directly related to the internal mental structure. Income is given from the outside, and the labor productivity is quite indirect even if skills are regarded as part of it. Game theory is more or less the same. |
| Shinzuki | Individual-forming collectivism considers the individual and society to be formed simultaneously by affecting each other. However, this is impossible if the internal mental structure is predetermined. The utility function is predetermined in market equilibrium theory and game theory. |
| Majime | Can we regard a decision criterion as an element for the internal mental structure? |
| Shinzuki | Yes, the decision criterion can be seen as part of the internal mental structure of the individual, similar to beliefs/knowledge of the individual. The individual has acquired beliefs/knowledge as well as behavioral/moral criteria by having interactions with society. |
| Morimori | It still sounds like evolutionary game theory. Once I thought I understood, but now I would like to ask you again if evolutionary game theory or the learning theory in game theory address such problems. |
| Shinzuki | No, neither does. Evolutionary game theory considers no change in the internal mental structure of the individual within one generation. There, each individual player is identified with a gene and is expressed by a strategy. The distribution of the strategies in the population may change over generations. |
| Morimori | How about the learning theory in game theory? |
| Shinzuki | In fact, learning theory in game theory doesn't specify what social phenomena it targets. The problem was arising from a mathematical point of view. It is simply the theory of algorithms and their convergences. The theory starts with the as- |

|  |  |
|---|---|
|  | sumption that strategies change under certain rules, and the rest is a matter of a differential equation or a difference equation. But the theorists want to use a fancy title for the theory. |
| Morimori | I'm surprised to hear that learning theory isn't more than a theory of algorithms and convergences. |
|  | Economics has already existed for several hundred years, but has never considered the internal mental structure of the individual? |
| Shinzuki | Economics is not yet that old. It has been about 200 years since Adam Smith, and game theory has existed for about 60 years since Neumann and Morgenstern. The tradition of economics is to avoid the problem of the human mind. The famous economist, Hirofumi Uzawa, once said, "The problem of the mind has always been a taboo in economics". This tradition came from behaviorism in social sciences or psychology, and was strongly emphasized in America in the first half of the $20^{th}$ century. |
| Morimori | Is that true? |
| Shinzuki | Yes, this is also affected by an ism. |
|  | In economics and game theory, utility theory and its extension, subjective probability theory, are regarded as dealing with the internal mental structure of the individual. Personally, I don't think that Savage's subjective probability theory has much content[12]. Game theory mentions an information structure. This isn't an internal structure but rather an external one since it is about how information is received. |
| Morimori | What is wrong with Savage's subjective probability theory as a description of the internal mental structure? |
| Shinzuki | In this theory, the subjective utility function and the subjective probability measure are derived from the preference relation of the individual. The preference relation is assumed to be the starting point, and the theory doesn't talk about how the preference relation is emerging or functioning with experiences. Util- |

---

[12] Savage LJ (1954) The foundations of statistics. John Wiley and Sons, New York.

ity is a matter of taste and thus, it must be subjective. It may be a possible assumption that the individual preferences are innate to the individual. However, probability is a matter of experiences and thought, and it shouldn't be purely innate in the individual's mind.

Savage's theory gives certain axioms on the preference relation. Then it shows the representation theorem that a preference relation is represented as a real-valued function, which is decomposed into a "utility function" and a "probability measure".

Majime  I think you continue as follows: The representation theorem is more or less an equivalence theorem between the preference relation with those axioms and the existence of a utility function and a subjective probability measure. Savage's theory talks all about the conditions for the representation theorem. More recently, nonexpected utility theory has been talking about generalizations. However, this literature doesn't target the internal mental structure of the individual at all.

Is my understanding of your claim correct?

Shinzuki  Yes, I think it is. I'd like to add a few words.

Those axioms are just the mathematical properties of the preference relation but are not about an aggregation of experiences or thought. The theory needs to talk also about some interface between the external experiences and internal mental structure.

Morimori  Professor, is there no theory that addresses the internal mental structure of the individual?

Shinzuki  I won't say that such a theory doesn't exist.

Morimori  Mm... Professor, you have often said, "I don't understand the double negation and you should state your sentence in a positive form", don't you? In this case, do you mean some theories exist?

Shinzuki  Yes, there are some theories [13]. The theory of Turing machines is one. The most extreme one is Neumann's theory of self-reproducing automata[14].

I suppose you know the theory of Turing machines. It formulates a computation process in the human mind. The theory gives the boundary between the computable and noncomputable functions. An important achievement is the discovery of the universal Turing machine. It computes any computable function as far as a program for it is given to the universal machine. This has become the theoretical foundation of the present-day computer.

Morimori  Could you please also explain Neumann's automata?

Shinzuki  Neumann's theory starts with a two-dimensional plane with lattices and 29 kinds of simple nerve cells. Each nerve cell is written in a box in the two-dimensional plane. Small organs are constructed by using these cells. Then, by combining these small organs, big organs are made. Finally, by combining these big organs, a self-reproducing machine is formed. The term "self-reproducing" means that the machine will write down an exact clone of itself in the two-dimensional plane. Another amazing point of this theory is that the machine has the universal Turing machine as its brain.

Morimori  What does this machine do?

Shinzuki  Neumann's self-reproducing automaton only reproduces clones as its children. Those clones will reproduce their clones again.

---

[13] Broadly speaking, there are many studies on the internal mental structure of a human, for example, Ryle G (1949) The concept of mind. Hutchinson, London gave a philosophical study on mind, and Gardner H (1985) The mind's new science. Basic Books, New York gave a quite exhaustive interdisciplinary study on the internal mental structure.

[14] Von Neumann J (1966) Theory of self-reproducing automata. (Burks AW editing and completing) University of Illinois Press, Chicago.

With some revisions, each automaton can make any computation because it has the universal Turing machine[15].

Unfortunately, Neumann passed away before he could complete this theory. If Neumann had survived long enough, maybe by now the world would be occupied by many self-reproducing automata.

Majime  Sir, it seems to do nothing else other than reproducing itself.

Shinzuki  That's right. Neumann challenged the mystery that a living creature can reproduce its children with the same structure. I think that Neumann wanted to show that it is mathematically and physically possible to formulate such a structure. In his theory, no mysterious elements are involved.

If we modify his theory slightly so that, for example, self-reproduction is only possible when a machine meets other machines, kills them and eats their parts, then these machines would play an evolutionary game.

Morimori  It is awful that they eat each other to reproduce themselves.

By the way, these machines can compute anything if they are given a program because they are furnished with the universal Turing machine.

Shinzuki  Exactly. This fact is important. Philosophically speaking, this shows that it is possible to move from Descartes' dualism of body and mind to Hobbs' monism. In this way, the mind is taken as the function of the machine. I think this is one thing Neumann wanted to show.

Morimori  But a machine has no emotions.

Shinzuki  It's not difficult to think about machines with emotions. Emotions are anger, anxiety, worry, happiness, pleasure etc. Those are functions of our mind, and, in fact, control our mind. Emotions play an important role in our behavior. Also, they are needed for our rational thinking. For example, you may feel

---

[15] The description here is extremely simplified. Although the theory was the work by Neumann alone, it looks like a big science in which many scientists participated.

|  |  |
|---|---|
|  | anger if my explanation is logically inconsistent. Then, your anger increases the power of your logical thinking to modify my explanation to a consistent one. This is just one instance. Emotions are playing important roles even in our rational thinking.<br>When a machine is sophisticated but is facing complicated tasks, it would need some emotions such as anger and pleasure to perform better. |
| Majime | When the ability of a machine is bounded, emotions may help the machine perform better. |
| Shinzuki | I agree. In a very complicated situation with a lot of contingencies, we cannot give a complete program to the machine to prescribe some action to every contingency. A machine needs to find what it should do by itself. If it feels anger and anxiety, it increases the intelligence level to solve the problem or if it feels comfort, then it takes a rest. |
| Majime | Are you going to say that love is also a function of the machine? |
| Shinzuki | Of course, yes. Neumann worked on a self-reproducing automaton. In the case of humans, we are making a bisexual reproduction. Love is an emotion to enhance the reproductive behavior. From the viewpoint of mind-body monism, the machine named "Majime" worries about love because the machine has a motive for reproductive behavior. It is formed in our evolution. |
| Majime | You are having fun with other people's worries. |
| Shinzuki | Sorry. |
| Majime | It would be better to get back to the original subject. I thought we were talking about (5): individual-forming collectivism. |
| Shinzuki | Yes, you are right. We were discussing the internal mental structure of the individual. Only if the internal mental structure is explicitly considered, it would be possible to see the formation of the individual. |

## Scene 4  Internal mental structure of the individual

Morimori   According to Neumann, the internal mental structure is something like organs consisting of cells. How can you relate this to society?

Shinzuki   You are right. Neumann's goal was to construct a machine as a physical entity with the functions of self-reproduction and thinking. My target is to consider not such organs as hardware but their functions as software. The functions addressed in Neumann's theory are extremely limited, and even thought is limited to mere computation.

Morimori   The function of the stomach is to make food small, the smaller intestines digest and absorb it, the larger intestines excrete it.

Majime   Here you go again. I'm not taken in a second time.
I think we should consider the mind as software.

Shinzuki   Indeed, for us the mind as software is more important than the brain as hardware. The mind has the functions of emotions and reason. Emotions control the behavior and intelligence level of the human machine. Reason is the intelligence itself to understand the natural and social environments.

Emotion is closer to the functioning of hardware, while reason is closer to software. Here we consider various types of reason.

Morimori   Are there different levels of reason?

Shinzuki   That is true. The word "reason" sounds as if you are simply indicating the computation ability or the ability to make theoretical arguments. It is, however, not so simple.

When we speak of a rational player in game theory, it indicates that he understands the structure of the game completely and that he can make the necessary computation instantaneously. In this case, the term "rational" is only used as an ornament before the actual analysis.

Morimori   Is that so?

Shinzuki   The day before yesterday, we discussed one small example. Even if you can use freely the four rules $+, -, \times, \div$ of arithmetic and inequality comparisons $\leq$, but if you don't know the radical expression $\sqrt{\ }$, you couldn't think about the irrational

|  | number $p = (30 - 2\sqrt{51})/29$. Such a player can't even compute the Nash equilibrium of that 3-person game[16]. Unless the language contains the radical expression $\sqrt{\phantom{x}}$, he couldn't have that irrational number $p$ in his mind. This means that language has an important role in determining the range of thought. |
|---|---|
| Majime | Although this isn't related to the standard idea of complexity of computation, it is one aspect of reason or rationality. This means that reason and rationality have many aspects. |
| Morimori | It is the same as teaching irrational numbers to high school students. First, you have to teach them the concept of $\sqrt{\phantom{x}}$. |
| Shinzuki | Exactly. We have learned a lot through experiencing things in society and communicating to others. Education, of course, plays an important role as well. Thus, certain parts will be formed through experience in society or by communicating with others. |
| Majime | Beliefs/knowledge and behavioral/moral criteria are formed as the individual's internal mental structure. Then this is consistent with (2): ontological individualism but neither with (1): reductionist individualism nor (3): identity-predetermined individualism. |
| Shinzuki | That's right. |
| Majime | I don't think that the standard economics and game theory handle this problem. Are you going to do research on this problem as a game theoretical problem? There seems to be a very large distance before concrete research becomes possible. |
| Shinzuki | Indeed, there is a great distance. That is why I have been working on that problem for more than twenty years. Because economics and game theory deal with humans and society, they will run into this problem someday. Therefore, the sooner you start, the more you contribute for our understanding of humans and society. |

---

[16] Act 4, Scene 4.

| | |
|---|---|
| Majime | You are right. In our profession all over the world, however, I hardly see any people who have been working on that problem. |
| Shinzuki | Nobody has been doing research on this in a systematic way. Therefore I want you and Morimori to be involved in this research. The question is, though, if you are ready for it. |
| Majime | I shall, later, think about your recruiting of me for research on that problem.

Now, I should conclude our methodological discussion. The internal formation and development of the individual is linked to the development of society. Therefore, such research is classified into (5): individual-forming collectivism, but since the basic unit of society is still the individual, it is also classified into (2): ontological individualism. |
| Shinzuki | Indeed. |
| Morimori | Now I can describe where you want to go. Each player is given artificial intelligence and is made to play the social game. |
| Shinzuki | That is one of our goals. The players are given language as well as computational and logical abilities. We teach them the rules of social behavior, and then make them play. At some point they start playing by themselves, affecting and educating each other and forming their own identities. Accordingly, a new society will be formed. |
| Majime | Are you saying that society, which consists of artificial players, starts moving autonomously? Then, what kind of research can we conduct more concretely? |
| Shinzuki | We can do research on how a boundedly rational player behaves, for example, what kind of problems occur with a player with or without knowing the radical expression $\sqrt{\phantom{x}}$. Bounded aspects of rationality such as falsities and misunderstandings should be studied. These were discussed when we talked about the Konnyaku Mondo. As a more concrete example, I can think of discrimination and prejudices. This is a good or bad example of individual-forming collectivism. First, prejudice exists as a mental attitude, and then discrimination as behavior arises. In- |

dividuals rationalize their discriminatory behavior, and new prejudices are formed. Or conversely, discrimination exists without prejudices, but prejudices are formed by rationalization, and then discrimination is reinforced more. This is a vicious circle[17].

Majime    I would like to hear a more concrete research proposal. What theoretical structure do you want to add to the present game theory?

Shinzuki    I already explained it to you but let me be more specific.

First, beliefs/knowledge and behavioral/moral criteria exist all in the mind of the individual. Behavior and judgments are inferred from those basic beliefs. Such inferences are studied by epistemic logic. But the formation of basic beliefs/knowledge is beyond a problem of logic. For example, the individual forms his view on society through experiences. This view of society consists of the individual's basic belief/knowledge and behavioral/moral criteria. Furthermore, the individual's communication transmits his view on society to others and therefore, communication plays an extremely important role in the formation of the view on society for the individual[18]. After this view is formed, it becomes a matter of internal logic.

Majime    Perhaps, I should summarize today's discussion. Even the formation of the individual's view of society is related to interactions in society, and these determine in turn the outlook of society itself. Therefore, such a research is taking the position of individual-forming collectivism, and at the same time, we consider the human person as the basic unit for this research. Thus, it also is ontological individualism.

---

[17] Cf. Act 2 and also Kaneko M, Matsui A (1999) Inductive game theory: discrimination and prejudices. Journal of Public Economic Theory 1: 101-137.

[18] This type of research has been stated. See Kaneko M, Kline JJ (2003) Modeling a player's perspective, Part I: Info-memory protocols, Part II: inductive derivation of an individual view. Mimeo.

|          | This is the connection between (2): ontological individualism and (5): individual-forming collectivism. |
|----------|---|
| Shinzuki | That's right. The only problem is how far we can go with this research program. |
| Morimori | Can I ask something that has to do with the future? |
| Shinzuki | Of course, go ahead. |
| Morimori | Will the players in the future behave more freely and maybe egoistically? |
| Shinzuki | I think so. Players will behave and think egoistically, and those with poor computation abilities will improve these abilities, beat clever players, start having relationships with other players and procreate. |
| Morimori | Professor, you are nasty. But in fact, we humans are seen as players in society. So we think egoistically, discuss and learn things in a way beneficial for ourselves. Eventually a player might say he wants to go for a drink!<br><br>We have been discussing like this for four days continuously. I'm really tired. Can we go for a drink tonight? Players need to recharge. Let's have beer, eat meat or fish and vegetables, and recharge with new energy and extra vitamins. |
| Shinzuki | That is a good idea. Shall we go to the sake bar, "Taishou"? I love the traditional cooking there! Majime, will you join us? |
| Majime   | Yes, today I will come with you. But if we go slightly earlier, that Mexican restaurant nearby has the happy hour. The margaritas are half price. Let's steep ourselves in a stylish atmosphere first, and then we can go to that cheap sake bar you both love. Until then, I will be back to my work. |

[Shinzuki, Majime and Morimori leave the stage]

Narrator: The final conclusion seems to be that even though the players might have their own ideas, they converge for drinking parties in the end! Is this the final implication of individual-forming collectivism? Well, they all will enjoy a party after four days of discussions.

I wish I could continue this play forever, but it is time to end as everything has its end. Dear reader, it has been a great pleasure for me to be narrator in this play. Farewell.

# Epilogue

[The poet appears silently on the stage]
  Hurry up, hurry up, I want to know the contents, sings the albatross
  Hurry up, hurry up, I should know the contents, sings the crested ibis
  Once time has passed and one knows the contents
  Only the crow remains, without distinction between back and front
[The poet leaves the stage silently]

[Shinzuki, Majime and Morimori appear on the stage and turn to the audience]

Shinzuki   Thank you very much for listening for such a long time.
　　　　　I'm sure you all know who the albatross and the ibis are. I wonder who the crow is. I hope it is not me! Also, what could he mean by saying there is no distinction between the back and the front?
　　　　　Well, there are quite many things I didn't even touch upon during the discussions. In latter acts, I was getting used to this style of discussion with Majime and Morimori as well as with others. Now, I want to continue, but for the moment this will be the end. We will have to leave, until another opportunity comes about. I feel sad.

Majime　　Thank you very much, too. I have found my rhythm for the discussions. As Professor Shinzuki said, I also feel sad to leave you, dear audience, in one half of my mind, but in the other half of my mind, I'm happy to have reached an end, since I should finish my own research paper.

Morimori　Hi, dear audience, good evening. The poet called me an albatross, such a stupid bird. But I don't think he is really right. You saw how I participated in the discussion. Don't you think that is quite an achievement?
　　　　　By the way, the paper I mentioned before is almost ready and I'm thinking about submitting it to the *Journal of Theoretical Economics*. To be honest, when I mentioned I proved that theo-

rem, I had no idea of how to write and submit a paper to an academic journal. Mr. Majime corrected my paper quite a few times, and I must thank him since he was kind and patient. I really hope my paper will get accepted for that journal. Mr Majime told me it is quite hard to pass the review process.

Shinzuki   Morimori, please keep it short. The author K is eager to say a few words. Now, ladies and gentlemen, it is my pleasure to present Mr. K, who has constructed this intellectual world. Mr. K, please come to the stage.

[K, wearing sunglasses, appears on the stage looking very clumsy. He unfolds a piece of paper and is about to start reading]

Morimori   Mr. K, your hands are shaking, are you okay?

K   Don't say such a thing! Now I'm embarrassed and tensed more. De, de dear audience, I... I am greatly ... uh, uh, what does it say here?

Majime   It appears to continue like, "I am greatly appreciative for your listening ...", something like that?

K   I see, it is "appreciative". Usually, I don't use such a word. Actually, it's the first time I'm giving a speech in public. So I asked my secretary, Ms Hizuki, to write a draft for this speech. Now I'm so nervous I can't even read her writing.

[K finds some of the audience not being able to stop laughing]

Hey you, audience, what do you expect? I'm not an actor. How am I supposed to be able to give a good speech? Ok, I should forget giving a formal and pretentious speech. Instead, I should speak from my own mind. Then, of course, I must complain about those actors.

Hey man, Kurai Shinzuki, you did quite a bit of ad-lib with your lines. And the ibis, Toru Majime, I should or shouldn't thank you for correcting my beautiful original text. Contrary to those guys, Genki Morimori was cute in his innocence. I loved his albatross behavior.

Shinzuki   I would be sorry if my ad-lib was bad. In the beginning I was acting as was written in the script, but I followed gradually the logic of the plot to continue the discussion. It was truly uninten-

|  | tional for me to make up for parts that were lacking or to make other parts more logical. This amazed me since logic was so powerful. As a result, it wasn't too bad, was it? Contrary to Oedipus' case, I pursued logic, expecting the outcome would become better. |
|---|---|
| Majime | I also would be sorry if you dislike my corrections. But, the original text contained a lot of poor and childish language. I couldn't stop myself revising my part, using more accurate, adequate, appropriate and esthetic expressions. I think that the result is much better than the original. Such childish language was quite fit to Morimori, though. |
| Morimori | Mr. Majime, it is unfair to say so. I didn't change my lines, since that is my character. Don't you understand my good acting? |
| K | Stop quibbling. Majime, did you call your changes esthetic? Where did you get the nerve to change your lines without asking the author? Is that esthetic behavior? You don't have the slightest idea about arts. I have written this masterpiece of art to compete with Plato's "The Republic" or Mozart's "Concerto for Clarinet". |
| Majime | Sorry, I didn't hear it. Did you say a "masterpiece of art" or a "masterpiece of crap"? |
| K | Yes, it's funny, but it's not esthetic. You should try harder to make a joke.<br>To be honest, I think I can compete with Aristophanes' "The Cloud". But it defeats mine only at a few points. I couldn't write such great lines as he did, |

*"Socrates was observing the orbit and the rotation of the moon, while looking upon the sky and opening his mouth. At that very moment a blue lizard, hidden in the night's darkness, peed on him"* [1].

---

[1] Aristhophanes: The cloud. In: Four texts on Socrates. Translated by West TG, West GS, (1984). Cornell University press, Ithaca.

[K points out to the audience on the right]

> Hey you, pale-faced cucumber, at the right end! You are putting your hand up and down. Do you want to say anything to me?

[K pricks up his ears]

> What? You imagined Shinzuki as my spokesman but the guy turned out to be much more noble than me? Are you trying to say I am vulgar? Or you can't see the difference between my back and front?

Shinzuki  Hahaha, you could say so! Aha, Mr.K is the crow the poet spoke of.

K  What else did you expect? I'm a real man, and he is just a character in a play.

> Ah ... the cucumber is putting his hand in the air again. Do you have something else to say?

[K pricks his ears again]

> You want to know if it is possible that characters in a play change the logic as well as the lines so as to fit in the play? Well, these three guys here did it, so it must be possible. Just the fact that it happened implies that my original text follows naturally and inevitably, and that it was written well. Did you understand, you pale-faced eggplant?

Morimori  The cucumber becomes an eggplant!

Shinzuki  The author doesn't seem to fit in a public setting like this. Mr. K, would you please leave the stage now? Thank you very much.

[Shinzuki shuffles K of the stage]

Shinzuki  I apologize for him being such a vulgar man. Please accept my apologies.

Morimori  Professor, what does "vulgar" really mean?

[Majime says instantly]

Majime  It means "common".

Morimori  I see, that is now vulgar knowledge among us, isn't it?

Shinzuki  Aha, that is nice.

Well, it is about time to finish. Dear audience, I hope that some day we will meet again. Majime, Morimori, next time let's make the discussion even more interesting. Well, goodbye to you all.

[The three leave the stage looking sad and lonely]

## Acknowledgements

This book is based on the author's interactions with many people during his long experience as a researcher and teacher as well as in his everyday life. Positive and negative memories of these interactions flashing across his mind are reflected in the discussions. The author wishes to express his gratitude to all the people who gave him such epiphanies, both positive and negative.

The author would like to thank the following persons, who directly or indirectly were related to the creation of this book. First, words of appreciation go to the actors who performed Act 1 at the Decentralization Conference in 2000, namely Ken Terao, Yuichi Osawa, and Takako Fujiwara. Secondly, gratitude is extended to Midori Hirokawa who gave comments on each of the acts and Sakae Nakano who criticized, revised and corrected a lot of the text. Hizuki Moriwaki helped with the preparation of the manuscript. The author's academic advisor Professor Mitsuo Suzuki always gave honest comments. Finally, the author was encouraged by Professor Hukukane Nikaido, who passed away in August 2001. Three months before his death he gave valuable comments on Act 3.

Looking back upon the completed acts 1 to 5, the author vividly remembers his time in the laboratory of Professor Mitsuo Suzuki from 1971 till 1977, in the office of Professor Martin Shubik from 1980 till 1982 and the joint research meetings with Professor Masahiro Okuno from 1986 till 1989. All the discussions during these times with so many people come back to mind. In addition, the environment of his institute at the University of Tsukuba to which the author has belonged to from 1977 to date (being affiliated to other universities for six years) has influenced the author more than he might realize. If the institute had not been there, this book had not been possible.

Interlude 2 is based on conversations with Jeffrey J. Kline and Oliver Schulte, who visited our institute in December 2001. The author would like to express his gratitude to them.

The author is deeply indebted to various people to have this English version. First, Ruth Vanbaelen prepared a basic translation of each act of the book, and then together with the author has revised the translation many times. A lot of valuable, both editorial and substantive, comments on earlier translations by Jeffrey J. Kline and Oliver Schulte are greatly appreciated. Helpful comments by Radha Balkaransingh and Sawako Shirahase are appreciated. The author is also grateful for Nobuhide Maeoka for the beautiful illustrations.

# The author

Mamoru Kaneko was born in Tokyo in 1950. In 1977 he finished the doctoral program at Tokyo Institute of Technology and obtained a Ph.D. in information sciences in 1979. After being assistant professor at the University of Tsukuba and associate professor at Hitotsubashi University, he became professor at the Virginia Polytechnic Institute and State University. Presently he is professor at the Institute of Policy and Planning Sciences of the University of Tsukuba.

Druck: betz-druck GmbH, D-64291 Darmstadt
Verarbeitung: Buchbinderei Schäffer, D-67269 Grünstadt